SpringerBriefs in Electrical and Computer Engineering

Series editors

Woon-Seng Gan, School of Electrical and Electronic Engineering, Nanyang Technological University, Singapore, Singapore

C.-C. Jay Kuo, University of Southern California, Los Angeles, CA, USA

Thomas Fang Zheng, Research Institute of Information Technology, Tsinghua University, Beijing, China

Mauro Barni, Department of Information Engineering and Mathematics, University of Siena, Siena, Italy

SpringerBriefs present concise summaries of cutting-edge research and practical applications across a wide spectrum of fields. Featuring compact volumes of 50 to 125 pages, the series covers a range of content from professional to academic. Typical topics might include: timely report of state-of-the art analytical techniques, a bridge between new research results, as published in journal articles, and a contextual literature review, a snapshot of a hot or emerging topic, an in-depth case study or clinical example and a presentation of core concepts that students must understand in order to make independent contributions.

More information about this series at http://www.springer.com/series/10059

Fanggang Wang • Guoyu Ma

Massive Machine Type Communications

Multiple Access Schemes

 Springer

Fanggang Wang
State Key Laboratory of Rail Traffic Control
and Safety
Beijing Jiaotong University
Beijing, China

Guoyu Ma
State Key Laboratory of Rail Traffic Control
and Safety
Beijing Jiaotong University
Beijing, China

ISSN 2191-8112 ISSN 2191-8120 (electronic)
SpringerBriefs in Electrical and Computer Engineering
ISBN 978-3-030-13573-7 ISBN 978-3-030-13574-4 (eBook)
https://doi.org/10.1007/978-3-030-13574-4

Library of Congress Control Number: 2019934772

This Springer imprint is published by the registered company Springer Nature Switzerland AG.
The registered company address is: Gewerbestrasse 11, 6330 Cham, Switzerland

Preface

Currently, the thriving of the Internet of things (IoT) provokes the industry passion on machine-type communications (MTC). Different with traditional human-type communications, MTC in IoT focus on the requirements for connectivity and reliability. To this end, two emerging scenarios named massive machine-type communications (mMTC) and ultra-reliable low-latency communications (URLLC) are proposed in the next-generation communications system. Therein, mMTC is able to serve a large number of miniaturized sensors and actuators in the intelligent sensing and control systems, which is a key support for the future industrial revolution.

One main challenge in mMTC is how to support massive connections with limited radio resources and low energy consumption. In order to overcome this challenge, favorable multiple access schemes are of great importance. Even though connectivity is the main requirement of mMTC, reliable access is also needed due to the potential IoT applications in the future. However, conventional multiple access scheme cannot guarantee the high connectivity and reliability under the limited radio resources and low energy supply. Therefore, novel multiple access schemes have to be considered.

In this book, the mMTC system is briefly demonstrated, and the suitable random access procedure is discussed. Also, various multiple access schemes for grant-free random access in mMTC are outlined. The first chapter will provide a brief introduction on mMTC, which is a core scenario of the next-generation communications system. Therein, the requirements and characteristics of mMTC are described. Then, the random access procedure for mMTC will be introduced in Chap. 2. According to the existence of the coordinations between base station (BS) and user equipments (UE), the random access can be categorized into grant-based random access and grant-free random access. The mMTC system prefers grant-free random access for the sake of saving heavy control signaling overheads. However, without the aid of those overheads, reliable multiple access protocol becomes challenging, so novel multiple access schemes have to be considered. In the following three chapters, three emerging multiple access schemes for grant-free random access will be presented. Those schemes apply different approaches to not only realize the functions of the conventional multiple access scheme but also undertake the tasks that originally require signaling overheads. In the meantime, numerous simulation results will be given to reveal more insights quantitatively.

Beijing, China   Fanggang Wang
                                                                                  Guoyu Ma

Acknowledgments

This work was supported in part by the National Key R&D Program under Grants 2016YFE0200900 and 2016YFB1200102-04, in part by the National Natural Science Foundation under Grants 61571034 and U1834210, in part by the Beijing Natural Science Foundation under Grant 4182051, in part by the Beijing Natural Haidian Joint Fund under Grant L172020, in part by the State Key Laboratory of Rail Traffic Control and Safety under Grant RCS2018ZT016, in part by the Key Laboratory of Universal Wireless Communications (BUPT), Ministry of Education, P. R. China, under Grant KFKT-2018102, in part by the Major projects of Beijing Municipal Science and Technology Commission under Grant Z181100003218010, in part by the Fundamental Research Funds for the Central Universities under Grant 2018JBM078, and in part by Nokia. A very special thanks to Professor Sherman Shen who made this book possible.

Contents

Contents

Acronym

3GPP	3rd Generation Partnership Project
AWGN	Additive white Gaussian noise
BLER	Block error rate
BPSK	Binary phase-shift keying
BS	Base station
CDF	Cumulative distribution function
CDF	Probability density function
CDMA	Code-division multiple access
CSA	Coded slotted ALOHA
CSI	Channel state information
CSMUD	Compressive sensing-based multiuser detection
CTSMA	Coded tandem spreading multiple access
eMBB	Enhanced mobile broadband
EPA	Extended pedestrian A model
ETU	Extended typical urban model
EU	European Union
GOMP	Group orthogonal matching pursuit
IIoT	Industrial Internet of Things
IoT	Internet of Things
LS	Least square
LTE	Long-term evaluation
MAC	Medium access control
MAI	Multiple access interference
MBB	Mobile broadband
MDS	Maximum distance separable
mMTC	Massive Machine-type communications
MTC	Machine-type communications
NB-IoT	Narrowband IoT
OFDM	Orthogonal frequency-division multiplexing
OMA	Orthogonal multiple access scheme
OMP	Orthogonal matching pursuit

PLNC	Physical-layer network coding
QPSK	Quadrature phase-shift keying
RACH	Random access channel
RAN	Radio access network
RAR	Random access response
RS	Reed-Solomon code
SIC	Successive interference cancellation
SNR	Signal-to-noise ratio
TSNDMA	Tandem spreading network-coded division multiple access
UE	User equipment
URLLC	Ultra-reliable low-latency communications
ZC	Zadoff-Chu sequence

List of Figures

Chapter 1
Introduction on Massive Machine-Type Communications (mMTC)

With the rise of a new round of scientific and technological revolutions and industrial changes in the world, the emergence of a large number of new applications, new businesses, new fields and new markets pose more requirements and challenges to existing mobile communication technologies [1]. To this end, a cellular communication paradigm shift from the fourth-generation (4G) to the fifth-generation (5G) is provoked [2, 3]. In 2012, the European Union (EU) officially launched the mobile and wireless communications enables for the 2020 information society (METIS) project to conduct research on 5G mobile communication networks. In addition to METIS, the EU has launched a larger research project named 5G-PPP. Moreover, the UK government has established a 5G R&D center with a number of companies at Surrey University. South Korea has launched the "GIGA Korea" 5G project, and China has established IMT-2020 promotion group [4, 5]. As shown in Fig. 1.1, the next generation 5G system not only continues to promote the data rate based on mobile broadband (MBB) to yield enhanced mobile broadband (eMBB), but also proposes two new scenarios of massive machine-type communications (mMTC) and ultra-reliable low latency communications (URLLC). The introduction of the new scenarios reflects the innovation of 5G. Those emerging scenarios are diverse and comprehensive, providing support for the diversified industrial revolution in the future. mMTC and URLLC are both scenarios for machine-type communications (MTC), which supports the future applications in internet of things (IoT) industry such as industrial internet of things (IIoT) [6]. It is well known that IIoT is an important driving force for intelligent manufacturing in the future. In IIoT, URLLC enables ultra-reliable and low latency transmission for real-time automation control of industrial dynamic processes while mMTC provides connectivity for massive miniaturized industrial sensor or actuator type of devices.

As the core scenario of 5G, mMTC has attracted great attention from researchers in academia and industry [8, 9]. There are two main challenges to getting mMTC from proposal to reality. Obviously the first one is how to achieve its massive connections. The 3rd generation partnership project (3GPP) suggests connectivity

© The Author(s), under exclusive license to Springer Nature Switzerland AG 2019
F. Wang, G. Ma, *Massive Machine Type Communications*, SpringerBriefs in
Electrical and Computer Engineering, https://doi.org/10.1007/978-3-030-13574-4_1

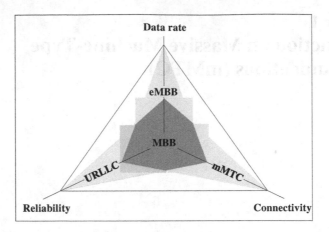

Fig. 1.1 Diverse scenarios and requirements in 5G [7]

of 10^6 devices/km^2 [10]. The transmission distance of mMTC determines that it cannot apply to high frequency band. In the meanwhile, the available low frequency band is scarce. The second one is the low energy consumption for each device in mMTC. In general, the service objects of mMTC system are the miniaturized, low-cost machine devices. Connecting a large number of devices to the grid is not practical so that each device is usually equipped with a battery which has limited life. If high energy consumption is allowed for each device, the batteries have to be changed frequently, which leads to a extreme low operation efficiency in the mMTC system.

In a mMTC system, massive machine devices are served by a single base station (BS) and the transmission period is dominated by the uplink transmission. Compare with human-type communications, the data packet size in mMTC is small [11]. At a point in time, only a subset of user equipments (UE) rather than the whole served UEs are transmitting simultaneously. The packet transmissions are highly dependent on the user activity. Consider a slot transmission based mMTC system with one BS and K UEs as shown in Fig. 1.2. Each UE has probability p_a to be independently activated to start the transmission in each time slot. The number of active UEs denoted by K_a follows the binomial distribution $\mathcal{B}(K, p_a)$ as

$$\Pr(K_a; K, p_a) = \binom{K}{K_a} p_a^{K_a}(1 - p_a)^{K-K_a} \tag{1.1}$$

which has the expected value as

$$\mathbb{E}(K_a) = Kp_a. \tag{1.2}$$

Fig. 1.2 Sporadic uplink transmission in the mMTC system [13]

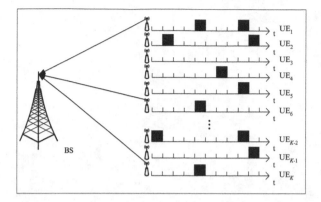

It is noted that the $K - K_a$ inactive UEs in a time slot still keep connecting with the BS rather than fall into completely sleeping. In order to describe this inactive state, 3GPP introduces a new state named RRC_INACTVE [12]. In the RRC_INACTVE state, the UEs do not transmit but still retains part of the radio access network (RAN) information and can quickly transit to the RRC_CONNECTED state through a paging-like message. With the involvement of RRC_INACTVE, the signaling overhead can be reduced and the energy consumption can be saved.

Due to short packet transmission, the number of active UEs determines the simultaneous packet arrivals in each slot. In addition, since the UEs are generally sensors or actuators, the data traffic in mMTC is intermittent. In other words, the active probability p_a is much less than 1 as $p_a \ll 1$ and the number of active UEs in each time slot is rare as

$$\mathbb{E}(K_a) \ll K. \tag{1.3}$$

This is named the sporadic characteristic of the mMTC system. Based on this characteristic, the mMTC system only needs to support simultaneous transmissions of rare UEs rather than all of the served UEs. This provides the chance for researchers to consider novel protocol to achieve massive connections with limited radio resources.

Chapter 2
Random Access Procedure for mMTC

In this chapter, the random access procedure for mMTC is discussed. First, the grant-based random access including the long term evolution (LTE) random access channel (RACH) is introduced. Then a suitable random access procedure for mMTC called grant-free random access is presented.

2.1 Grant-Based Random Access

In the grant-based random access, the UEs need to request the grant from the BS to schedule the transmissions. The LTE RACH is a typical grant-based random access. Currently, LTE RACH has been applied in narrowband IoT (NB-IoT), which is an emerging standardized technology for the IoT networks [14]. The LTE RACH is divided into non-contention-based random access and contention-based random access [15]. The non-contention-based random access procedure has three steps as shown in Fig. 2.1. In the first step, the BS assigns random access preambles to all the UEs. Then the each UE transmits the assigned preamble to the BS. After receiving the preambles, the BS sends the random access response (RAR) to all the UEs. Since each UE is assigned with a unique preamble, the collision in this procedure can be avoided. The contention-based random access procedure has four steps as shown in Fig. 2.2. At first, each UE transmits the random access preamble randomly selected from the pseudo-random sequences which are broadcast by the BS ahead of the random access period. It has the possibility that multiple UEs select the same preamble and the collision is inevitable. When receiving the preambles, the BS sends RAR to schedule the transmissions for the UEs whose preambles are successfully identified. Then the scheduled UEs starts the initial layer 3 message (Msg3) transmission. At last, the contention resolution is achieved by the BS in the downlink transmission.

F. Wang, G. Ma, *Massive Machine Type Communications*, SpringerBriefs in Electrical and Computer Engineering, https://doi.org/10.1007/978-3-030-13574-4_2

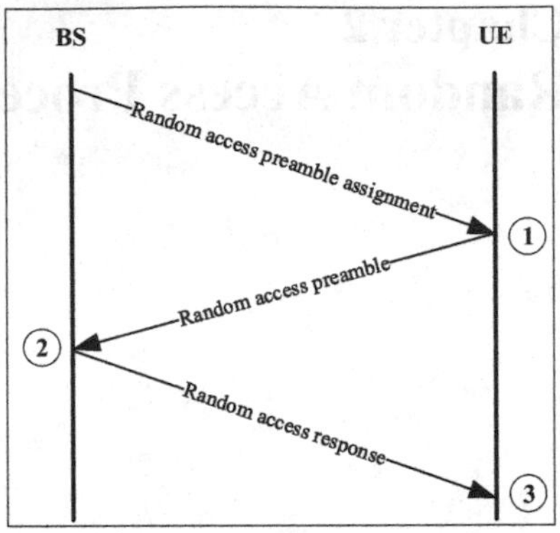

Fig. 2.1 Non-contention based random access

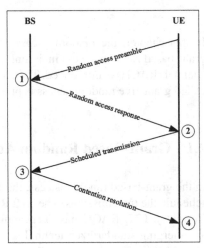

Fig. 2.2 Contention-based random access

In mMTC, the uplink transmission is dominant and the random access is usually initiated by the UEs. Thus contention-based random access can be considered as the benchmark protocol for the mMTC system.

2.2 Grant-Free Random Access

The grant-based random access is only designed for the small number of UEs. When facing with massive machine-type devices, the iterations between UE and BS such as the four step handshakes in the contention-based random access will result in tremendous control signaling overheads [16]. Currently, researchers are considering how to reduce the number of interactions between UE and BS in the mMTC system as much as possible. To this end, grant-free is introduced. In grant-free random access, the UE does not need the base station to perform dynamic and explicit scheduling authorization, but performs an "arrive-and-go" transmission

Fig. 2.3 Grant-free random access

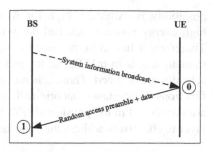

mode. Once the data arrives at one UE, the UE immediately transmits the data in the next transmission slot. As shown in Fig. 2.3, the grant-free random access is expected to be achieved within two steps including the system information broadcast. Compared to the grant-based random access, grant-free random access has the following advantages for mMTC: (1) because there is no or little need for scheduling, signaling overhead and transmission delay are reduced; (2) Since it can be sent the data packet immediately, the UEs only need to be activated when there is a transmission request, so that the UE can have higher energy efficiency; (3) The UE can be designed at a very low cost, and the computational complexity is mainly undertaken by the BS. However, the use of grant-free random access also has its corresponding challenges. First, due to the lack of coordination between BS and UE, the data packets received by the BS are anonymous. Thus user identification is required to be implemented by the BS to distinguish the active UEs from a large number of served UEs. Second, since resource allocation and scheduling for massive UEs are omitted, multiple UEs possibly occupy the same channel resources. If these UEs are activated and transmit data in the same time slot, they will collide at the base station to degrade the access reliability.

The ALOHA protocol can be categorized into the grant-free random access [17]. The ALOHA is divided into pure ALOHA and slotted ALOHA. In pure ALOHA, when the UE has data to transmit, it immediately transmits the data packet to the BS. The collision occurs when more than one data packets simultaneously arrive at the BS. The colliding UEs will stop for a while and attempt the transmission again. Slotted ALOHA is the improvement to the pure ALOHA protocol. The idea is to use the clock to unify the user data transmission. In slotted ALOHA, the uplink transmission period is divided into multiple time slots. Each transmission can only be triggered at the beginning of a slot and the UE must wait until the next time slot to start sending data. The data packet size should be less than or equal to the time slot length. Compare to pure ALOHA, the randomness of data transmission in slotted ALOHA is alleviated so that the collision possibility is greatly reduced. In the aforementioned contention-based random access procedure, the randomly selected preambles are transmitted based on slotted ALOHA.

Nevertheless, the existing challenges of grant-free random access cannot be appropriately tackled by the ALOHA protocol in a mMTC system. Since the scarce radio resources are occupied by the massive uncoordinated UEs, the collision possibility is extremely high for ALOHA. High collision probability results in high energy consumption and delay to significantly degrade the mMTC system. Therefore, it has to be re-consider the multiple access scheme for the grant-free random access in mMTC. In the rest of this chapter, three novel multiple access schemes are outlined. Those schemes realize both the conventional multiple access functions and the tasks that originally require signaling overheads. Particularly, they are introduced in the physical (PHY) layer and the medium access control (MAC) layer to effectively address the collision problem.

2.3 Summary

In this chapter, two random access procedures including grant-based random access and grant-free random access have been introduced. It has been shown that grant-free is preferred by the mMTC system for saving the huge control signaling overheads.

Chapter 3
Compressive Sensing Based Multi-user Detection (CSMUD)

Compressive sensing based multi-user detection (CSMUD) is a code domain PHY layer solution which attracts a lot of attentions [18, 19]. CSMUD takes advantage of the sparse user activity of the mMTC sporadic transmission for user identification and data detection. In CSMUD, compressive sensing theory is effectively applied to alleviate the influence caused by the cross-correlation between non-orthogonal spreading sequence on user identification and data detection.

3.1 System Model

The block diagram of CSMUD is shown in Fig. 3.1. Suppose a single cell system model with one BS and K UEs which transmit in the slot manner. Each UE is activated with probability p_a in each slot to transmit the channel coded bits. For an active UE k, after channel encoding, modulation is implemented to map the bits into the alphabet $\mathcal{A}$ as $\mathbf{d}_k \in \mathcal{A}^{b \times 1}$. In this chapter, binary phase shift keying (BPSK) modulation is assumed for simplicity so that $\mathcal{A} = \{-1, 1\}$. Afterwards, the b data symbols are spread by the sequence $\mathbf{s}_k \in \mathbb{C}^{q \times 1}$ as

$$\mathbf{x}_k = \mathbf{S}_k \mathbf{d}_k \tag{3.1}$$

where $\mathbf{x}_k \in \mathbb{C}^{qb \times 1}$ and q is the spreading factor and $\mathbb{C}$ denotes the set of complex numbers. The spreading matrix $\mathbf{S}_k \in \mathbb{C}^{qb \times b}$ is denoted as

$$\mathbf{S}_k = \begin{bmatrix} \mathbf{s}_k & & & \\ & \mathbf{s}_k & & \\ & & \ddots & \\ & & & \mathbf{s}_k \end{bmatrix}. \tag{3.2}$$

F. Wang, G. Ma, *Massive Machine Type Communications*, SpringerBriefs in Electrical and Computer Engineering, https://doi.org/10.1007/978-3-030-13574-4_3

9

Fig. 3.1 Block diagram of CSMUD

Here the spreading sequence is pseudo-noise (PN) sequence such as the real-valued PN $\mathbf{s}_k \in \{-1, 1\}^{q \times 1}$. In CSMUD, each UE in the system has a unique spreading sequence so that the BS is able to identify the active UEs from the received signal. If UE k is inactive, the transmitted signal can be represented by the zeros symbols as $\mathbf{d}_k = \mathbf{0}$. Hence, augment alphabet $\mathcal{A}_0 = \{-1, 0, 1\}$ is introduced to cover both active and inactive UEs.

Assume the signals from all the active UEs in a slot are received synchronously. The synthesized signal at the BS receiver is written as

$$\mathbf{y} = \sum_{k=1}^{K} \mathbf{H}_k \mathbf{S}_k \mathbf{d}_k + \mathbf{n} \tag{3.3}$$

where $\mathbf{H}_k$ is the convolutional matrix of the L_h-length channel impulse response $\mathbf{h}_k \in \left[h_{k,0}, h_{k,1}, \ldots, h_{k,L_h-1} \right]^T$. $\mathbf{H}_k$ is written as

$$\mathbf{H}_k = \begin{bmatrix} h_{k,0} & 0 & \cdots & 0 \\ h_{k,1} & h_{k,0} & \ddots & \vdots \\ \vdots & h_{k,1} & h_{k,0} & 0 \\ h_{k,L_h-1} & \vdots & h_{k,1} & \ddots \\ 0 & h_{k,L_h-1} & \vdots & \ddots \\ \vdots & \vdots & h_{k,L_h-1} & \vdots \\ 0 & 0 & \cdots & \ddots \end{bmatrix}. \tag{3.4}$$

For simplicity, the tail of the received signal is not captured. Then the last $L_h - 1$ rows of the convolution matrix is deleted and $\mathbf{H}_k$ becomes a $qb \times qb$ matrix. Subsequently, Eq. (3.3) can be rewritten in the vector as

$$\mathbf{y} = \mathbf{A}\mathbf{x} + \mathbf{n} \tag{3.5}$$

where $\mathbf{x} \in \mathbb{C}^{Kb \times 1}$ is

$$\mathbf{x} = \left[\mathbf{d}_1^T, \mathbf{d}_2^T, \ldots, \mathbf{d}_K^T \right]^T \tag{3.6}$$

and $\mathbf{A} \in \mathbb{C}^{qb \times Kb}$ is

$$\mathbf{A} = [\mathbf{H}_1 \mathbf{S}_1, \mathbf{H}_2 \mathbf{S}_2, \ldots, \mathbf{H}_K \mathbf{S}_K]. \tag{3.7}$$

It can be found that if the k-th UE is inactive, the corresponding qb entries in $\mathbf{x}$ are zeros so that $\mathbf{A}$ and $\mathbf{x}$ can be regarded as block sparse. For block sparsity, a compressive sensing theory based algorithm called group orthogonal matching pursuit (GOMP) is considered for both user identification and data detection in CSMUD.

3.2 Group Orthogonal Matching Pursuit (GOMP) Algorithm

As shown in Fig. 3.2, the GOMP algorithm is an extension of the orthogonal matching pursuit (OMP) algorithm for block sparsity. In the OMP algorithm, the receiver correlates each column of the matrix $\mathbf{A}$ to explore the sparsity of each symbol for each UE. But in the GOMP algorithm, the receiver directly correlate all the columns belonging to a UE in $\mathbf{A}$ instead of correlating each individual column

Fig. 3.2 The GOMP algorithm

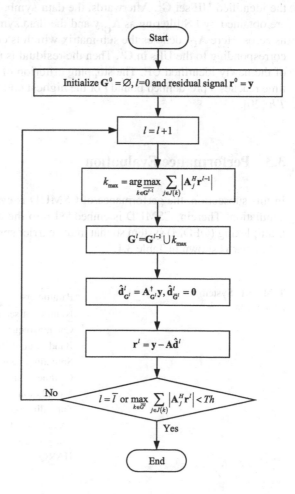

to fully exploit the block sparsity. Assume the channel state information (CSI) is perfectly known at the receiver. The correlation for UE k can be written as

$$\omega_k = \sum_{j \in J(k)} \left| \mathbf{A}_j^H \mathbf{y} \right|. \tag{3.8}$$

Here $J(k)$ denotes the column indices of UE k, $(.)^H$ denotes the Hermitian transpose and $|.|$ indicates the absolute value function. In GOMP algorithm, the K_a UEs with the highest correlations ω_k are obtained to achieve the user identification. In the meanwhile, least square (LS) filtering is implemented to detect the data of each identified UE. The GOMP algorithm is shown below.

In this algorithm, the identified UE group $\mathbf{G}^0$ is initialized as the empty set, the iteration $l = 0$ and the residual signal is set as $\mathbf{r}^0 = \mathbf{y}$. In each iteration, the receiver searches the residual UE set $\bar{\mathbf{G}}^{l-1}$ to seek the UE with the highest correlation to the residual signal. The UE $k_{\max}$ who has the highest correlation is classified into the identified UE set $\mathbf{G}^l$. Afterwards, the data symbols of the identified UEs in $\mathbf{G}^l$ are obtained by LS filtering as $\mathbf{A}_{\mathbf{G}^l}^{\dagger} \mathbf{y}$ and the data symbols of the UEs in $\bar{\mathbf{G}}^l$ are set as zeros. Here $\mathbf{A}_{\mathbf{G}^l}$ denotes the sub-matrix which is composed of the columns in $\mathbf{A}$ corresponding to the UEs in $\mathbf{G}^l$. Then the residual is updated by deleting the signal of the newly identified UE. The stopping criterion of this algorithm is the iteration time reaches a pre-defined bound $\bar{l}$ or the highest correlation is less than a threshold Th [20].

3.3 Performance Evaluation

In this subsection, the performance of CSMUD is evaluated through the link-level simulation. Therein, CSMUD is embedded into the orthogonal frequency division multiplexing (OFDM) system so that multi-carrier spreading is applied. The system parameter is shown in Table 3.1.

Table 3.1 System parameters

Parameters	Value
Number of served UEs	52
Carrier frequency (GHz)	2
Bandwidth (kHz)	195
Spreading factor	13
Channel coding rate	1/2
Modulation	QPSK
Spreading sequence	Zadoff-Chu sequence
Channel model	ETU, EPA
Channel estimation	Ideal
HARQ	No

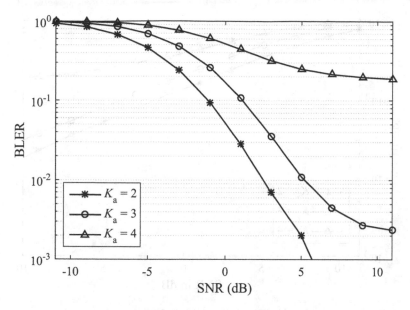

Fig. 3.3 BLER performance with different K_as under the ETU channel

The simulation uses 13 subcarriers to transmit data and each subcarrier occupies a bandwidth of 15 kHz so that the total bandwidth is 195 kHz. The adaptive modulation coding is not set in this simulation. The modulation is fixed to quadrature phase shift keying (QPSK) and Turbo coding with code rate 1/2 is used. In addition, the Zadoff-Chu sequence (ZC) is used for spreading [21]. Since prime length spreading sequences have good orthogonality, the spreading factor in this simulation is set to $q = 13$. In order to serve all the UEs, 4 different ZC roots are applied and each root generates 13 ZC sequences by cyclic shifting. The cross-correlation is 0 for the sequences with identical root and $\sqrt{q}$ for the sequences with different roots.

Figure 3.3 shows the block error rate (BLER) of CSMUD with different active UE number K_a under the extended typical urban model (ETU) channel. It is noted that the maximum delay of the ETU channel is 5 μs [22]. The channel coherence bandwidth is larger than the total system bandwidth so that the channel can be regarded as a flat fading channel. As can be seen from the figure, when the $K_a = 2$, the system has good detection performance. When the active UE number increases to 3, it can be found that the system performance has the error floor in high signal-to-noise ratio (SNR). This is because when 3 UEs simultaneously send data, the CSMUD has the error in user identification. This error mainly comes from the cross-correlation between the spreading sequences with different ZC roots. The cross-correlation results in interference on user identification. If the number of active UEs is increased to 4, the detection performance degrades greatly and the error floor occurs earlier. Based on above, it can be obtained that the detection performance of CSMUD is mainly influenced by the sequence cross-correlation which is determined by the active UE number.

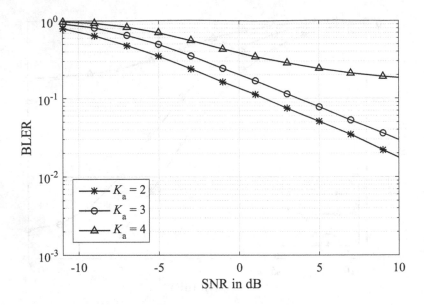

Fig. 3.4 BLER performance with different K_as under the EPA channel

Figure 3.4 shows the BLER performance with different K_as under the extended pedestrian A model (EPA) channel. The maximum delay of the EPA channel is 410 ns so that the channel is still flat fading. It can be observed from the figure that regardless of the number of active UEs, the detection performance under the EPA channel is worse than that under the ETU channel. When $K_a = 4$, the detection performance is still a significant drop compare to the cases for $K_a = 2$ and 3. Also, there is an error floor for $K_a = 4$ due to the cross-correlation. However, unlike the ETU channel, the curves for $K_a = 2$ and 3 do not appear flat.

Figure 3.5 provides the BLER performance of with different UE speeds v under the EPA channel. One can get from this figure that when $v < 30$ km/h, the moving speed has a limited influence on the detection performance of CSMUD. But when the speed increases to 100 km/h, the performance tends to be flat in the high SNR region due to the error in user identification. Therefore, the current CSMUD version is not suitable for high speed mobile applications.

According to the results above, it can be concluded that CSMUD is able to reconstruct the signals of 2 and 3 colliding UEs. For more active UEs, the performance has a significant deterioration caused by the sequence cross-correlation. With a large number of served UEs, if K_a is not small, the user activity is no longer sparse. The user activity sparsity should keep an inverse proportional relationship to the number of served UEs K. Hence, there exists a tradeoff between the connectivity and the spectral efficiency represented by the active UE number. In order to fully exploit CSMUD, the related applications in IoT should properly set the number of connections and the UE active probability based on their own business feature.

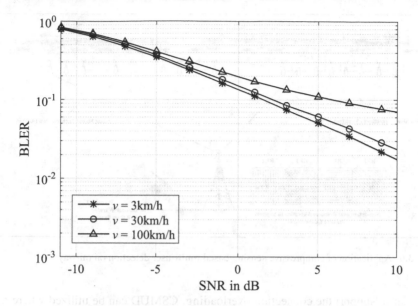

Fig. 3.5 BLER performance with different UE speeds v under the EPA channel

At last, we introduce the application of CSMUD on the future railway system. Railway is considered as a key scenario in the future 5G era. Nowadays, various emerging applications are proposed such as critical high-definition video surveillance, high reliability train control signaling service and IoT for railways [23]. Therefore, the railway also needs to face the challenges for higher spectral efficiency, reliability and connectivity [24]. Particular, for the sake of saving human cost on the maintenance of railway devices, the IoT applications on the railway scenario gain plenty of interests from the communication market. In this section, a sensing system is presented to discuss the massive connectivity in the railway system.

Figure 3.6 shows a sensing system for the disaster warning and rail flaw alarm. In this system, massive sensor devices with single antenna which are allocated along the rails are connecting with a singe BS with single antenna. If there has no alarm, the sensor is awaiting and keeps connection with the BS. If an alarm occurs and detected by the sensor, the transmitter of the device immediately sends the accident signal to the BS at the warning control room. In order to effectively receive the warning signal in time, the BS is required to keep connections with all the massive sensors. Therefore, the identification of each transmitting sensor is necessary. Nonetheless, the radio resource provisions in railway applications cannot support the orthogonal identifications of those sensors. Thus non-orthogonal resource allocation has to be utilized to support overloaded connections. In the meanwhile, due to the occurrence of accidents such as disasters and rail flaws, the transmissions of accident signals can be regarded as sporadic. In addition, the probability of multiple activations is much less than the single activation. Hence,

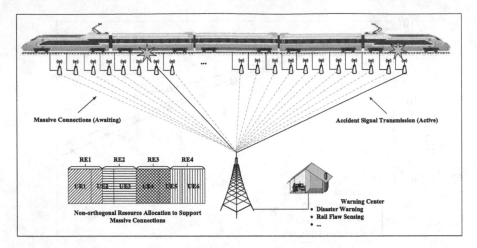

Fig. 3.6 Application of compressive sensing based multi-user detection on railway

in order to support the connection overloading, CSMUD can be utilized where the sparsity of sensor activity can be exploited. Specifically, user identification for the massive connected sensor devices can be achieved in the single activation situation.

It can be seen that CSMUD is suitable in this railway application. However, there still have some challenges need to be considered. One is the near far effect. If TDD is applied, then the large scale fading of the sensor cannot be well compensated to result in the near far effect. Generally, with near far effect, SIC can be more effective. But the channel states of the active users should be reported to the BS for SIC. Without the knowledge of channel states, SIC is imperfect and error propagation exists.

3.4   Summary

In this chapter, the scheme CSMUD has been presented which takes advantage of the sparse user activity of the mMTC sporadic transmission for user identification and data detection. Through numerical results, it has been shown that the influence of MAI on user identification and data detection can be effectively alleviated via utilization of the compressive sensing theory.

Chapter 4
Coded Slotted ALOHA (CSA)

In this chapter, a MAC layer scheme named coded slotted ALOHA (CSA) is presented with illustrative example and theoretical analysis [25, 26]. The difference between the PHY layer schemes and the MAC layer schemes is that the PHY layer solutions are used for the single slot transmission process while the MAC layer solutions are used for multi-slot transmission process. In CSA, each served UE selects multiple time slots to transmit the replica of its data packet and successive interference cancellation (SIC) is implemented at the receiver to resolve the collision to realize high connectivity.

4.1 CSA System Description

Figure 4.1 shows the CSA system with one BS and K UEs who transmit in the uplink contention period with T time slots. In CSA, each UE randomly selects multiple slots from the uplink contention period to transmit the replicas of its data packet. In the meantime, each UE adds a pointer on each of its replicas to indicate the positions of other replicas [27]. The probability that a slot is selected by an UE is p_a so that the selected slot number T_u for each UE follows the binomial distribution $\mathcal{B}(T, p_a)$ as

$$\Pr(T_u; T, p_a) = \binom{T}{T_u} p_a^{T_u} (1 - p_a)^{T - T_u}. \tag{4.1}$$

Similarly, active UE number K_a in each slot follows the binomial distribution $\mathcal{B}(K, p_a)$ as shown in Eq. (1.1).

At the receiver side, the signals from all the slots in the uplink contention period are detected to find the collision free slots which are only occupied by one active UE. The slot is called the singleton slot where the occupied UE is decodable. After

© The Author(s), under exclusive license to Springer Nature Switzerland AG 2019   17
F. Wang, G. Ma, *Massive Machine Type Communications*, SpringerBriefs in
Electrical and Computer Engineering, https://doi.org/10.1007/978-3-030-13574-4_4

Fig. 4.1 Coded slotted
ALOHA system

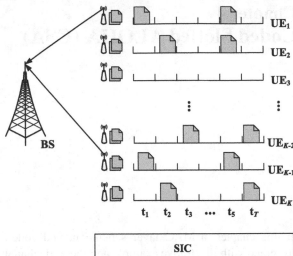

Fig. 4.2 SIC at the receiver
of CSA

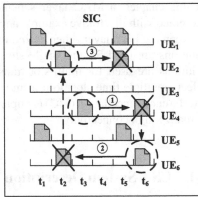

decoding the data packet of the UE at the singleton slot, the receiver is able to know
the slot positions of other replicas through the pointer. Then SIC is implemented
to remove those replicas of this active UE and some of the corresponding slots are
possible to become collision free. Therefore, SIC is iteratively executed until there
has no singleton slot. One example is shown in Fig. 4.2. In this example, at first only
UE_4 is collision free at slot t_3. Then the data packet of this UE is decoded and the
replica at slot t_6 can be removed. After removing the replica, t_6 becomes a singleton
slot and UE_6 is able to be decoded. Likewise, the replica of UE_6 at t_2 is deleted and
UE_2 is decodable. At last, even though the replica of UE_2 at t_5 can be removed, the
slot t_5 is still occupied by two residual UEs. In this manner, UE_2, UE_4 and UE_6
can be successfully received and re-transmissions are required for other UEs in this
example.

4.2 Theoretical Analysis of CSA

This subsection theoretically analyzes the performance of CSA. For one SIC iteration of CSA, the data packet replica of one active UE at a slot cannot be removed as long as the replicas in other slots cannot be decoded in the previous SIC iteration. According to [28], the average probability that a replica in one slot cannot be decoded is denoted as $P_{err,t}$. Thus for an active UE who has T_u replicas, the probability that a replica in one slot cannot be removed is shown as

$$P_{err,u}(T_u) = P_{err,t}^{T_u-1}. \tag{4.2}$$

Accordingly, the average probability of the replica removal is written as

$$P_{err,u} = \sum_{T_u=1}^{T} \Lambda_{T_u} P_{err,t}^{T_u-1} \tag{4.3}$$

where Λ_{T_u} is the probability that the replica is affiliated to the active UE who selects T_u slots for transmission. According to [29], this probability can be expressed as

$$\Lambda_{T_u} = \frac{T_u \Pr(T_u; T, p_a)}{\mathbb{E}[T_u]} = \frac{T_u \Pr(T_u; T, p_a)}{T p_a}. \tag{4.4}$$

By combining Eqs. (4.3) and (4.4), $P_{err,u}$ is re-written as

$$P_{err,u} = \sum_{T_u=1}^{T} \frac{T_u \Pr(T_u; T, p_a)}{T p_a} P_{err,t}^{T_u-1}. \tag{4.5}$$

Subsequently, for a single slot, consider the data packet replica of one active UE is decodable if $K_a - 1 - K_{a,re}$ UEs have been removed. Thereby, for a slot with K_a active UEs, the probability that a replica can be decoded is shown as

$$1 - P_{err,t}(K_a) = \sum_{K_{a,re}=0}^{K_a-1} \mathcal{P}(K_{a,re}) \binom{K_a-1}{K_{a,re}} (1 - P_{err,u})^{K_a-1-K_{a,re}} P_{err,u}^{K_{a,re}}. \tag{4.6}$$

Here $K_{a,re}$ is the number of remaining active UEs in the slot except the affiliated UE, and $\mathcal{P}(K_{a,re})$ is the probability that the replica can be decoded when the number of other remaining UEs is $K_{a,re}$. By averaging $1 - P_{err,t}(K_a)$ through different K_as, the average probability that one replica cannot be decoded in a slot can be written as

$$P_{err,t} = 1 - \sum_{K_a=1}^{K} \Omega_{K_a} (1 - P_{err,t}(K_a)) \tag{4.7}$$

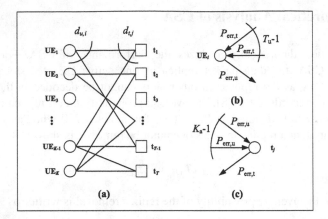

Fig. 4.3 Bipartite graph representation and the and-or tree evaluation of CSA

where Ω_{K_a} is the probability that the affiliated UE is at the slot which are occupied by K_a active UEs. This probability is described as

$$\Omega_{K_a} = \frac{K_a \Pr(K_a; K, p_a)}{\mathbb{E}[K_a]} = \frac{K_a \Pr(K_a; K, p_a)}{K p_a}. \tag{4.8}$$

Here $\Pr(K_a; K, p_a)$ indicates the binomial probability design function (PDF) of K_a with parameters K and p_a.

For the ease of analysis, the SIC process of CSA in the uplink contention period can be described by the bipartite graph as shown in Fig. 4.3a. In the bipartite graph, the circle nodes represent the K served UEs and the square nodes represent the T time slots. The edge connected between a user node and a slot node denotes that the UE is transmitting on this slot. In CSA, user degrees $d_{u,i}$ and slot degrees $d_{t,j}$ are applied to denote the number of edges connected to user node UE_i and slot node user node t_j respectively [26]. In other words, $d_{u,i}$ and $d_{t,j}$ indicate the number of slots selected by UE_i and the number of active UEs at slot t_j.

According to [28], and-or tree is introduced to provide an asymptotic performance analysis for CSA under a large K. Here $P_{err,t}$ denotes the average probability that one incoming edge to an user node cannot be decoded in the previous SIC iteration, and $P_{err,u}$ denotes the average probability that one incoming edge to an slot node cannot be decoded. Then the and-or tree evaluation is performed in an iterative manner to represent the SIC iterations. For the iteration $l \geqslant 1$, $P_{err,t}$ and $P_{err,u}$ can be derived as

$$P_{err,u}^l = f\left(P_{err,t}^{l-1}\right) \text{ and } P_{err,t}^l = g\left(P_{err,u}^l\right) \tag{4.9}$$

where $P_{err,u}^l$ and $P_{err,t}^l$ are the derivations of $P_{err,u}$ and $P_{err,t}$ in the l-th iteration. It is initialized that $P_{err,t}^0 = 1$. The functions $f(.)$ and $g(.)$ are depicted by Eqs. (4.5) and (4.7). Finally, the probability that the data packet of one active UE can be successfully transmitted in the uplink contention period is expressed as

$$P_D = 1 - \lim_{l \to \infty} P_{err,t}^l. \tag{4.10}$$

The transmission reliability can be represented by P_D. Hence, the relationship between the reliability, connectivity and data rate can be derived.

4.3 Summary

This chapter has described the MAC layer scheme CSA with illustrative example and theoretical analysis. In CSA, the collision can be resolved though data replica and SIC so that reliable access under high connectivity can be achieved.

Chapter 5
Tandem Spreading Multiple Access

Currently, high connectivity in the mMTC system can be achieved by the cutting-edge multiple access approaches. Nevertheless, a comprehensive multiple access scheme to cover both high connectivity and reliability in grant-free random access system is potentially demanded by the future MTC applications. Motivated by this, a novel spreading based multiple access technique named *tandem spreading multiple access* has been proposed. This approach is initiated from the scheme called *tandem spreading network-coded division multiple access* (TSNDMA) [30]. Then a generalized version of TSNDMA named *coded tandem spreading multiple access* (CTSMA) is introduced [31]. In the following subsections, those two versions are presented. In addition, the multi-slot design of CTSMA is described [32].

5.1 Tandem Spreading Network-Coded Division Multiple Access (TSNDMA)

Figure 5.1 shows the block diagram of TSNDMA. Compare to the conventional transmission system, two new procedures named physical layer network coding (PLNC) and tandem spreading are involved. In TSNDMA, multiple spreading sequences are utilized to spread the data bits of one UE. The bits after channel encoding are divided into segments. Then a segment-specific spreading sequence is utilized to spread the data symbols of each segment. Here the tandem spreading combination is generated by combining the sequences corresponding to all the segments. Each user equips a unique tandem spreading combination and the tandem spreading combinations for all the served UEs constitute the tandem spreading codebook. Therefore, user identification in TSNDMA can be achieved by recognizing the combinations from tandem spreading codebook. However, if the spreading sequences at the same segment position of the combinations for two active users are same, the collision occurs to influence the data detection performance.

© The Author(s), under exclusive license to Springer Nature Switzerland AG 2019 23
F. Wang, G. Ma, *Massive Machine Type Communications*, SpringerBriefs in
Electrical and Computer Engineering, https://doi.org/10.1007/978-3-030-13574-4_5

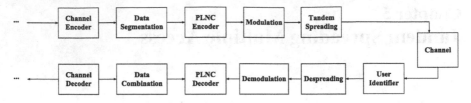

Fig. 5.1 Block diagram of TSNDMA

To this end, TSNDMA employs PLNC is to generate a redundancy segment for collision recovery.

In TSNDMA, the channel coded bits of user k who is active are denoted as $\mathbf{d}_k \in \{0, 1\}^{b \times 1}$. Then the bit vector $\mathbf{d}_k$ is divided into m segments as $\mathbf{d}_k^i$ for $i \in [1, m]$. If $\frac{b}{m}$ is not an integer, then some extra bits are appended so that each segment has the size $b_m = \lceil \frac{b}{m} \rceil$ where $\lceil . \rceil$ is the ceiling function. After data segmentation, PLNC encoding is implemented by adding all the m bit vectors in modulo 2 to generate the redundancy segment as

$$\mathbf{d}_k^n = \sum_{i=1}^{m} \mathbf{d}_k^i \mod (2) \tag{5.1}$$

where $n = m + 1$ and $\mathbf{d}_k^{m+1}$ is the bit vector of the redundancy segment. After PLNC encoding, b_m bits of the k-th user at segment i for $i \in [1, n]$ are modulated to the data symbols as $\mathbf{x}_k^i \in \mathcal{A}_0^{b_m \times 1}$. Similar to the aforementioned CSMUD, the augment alphabet $\mathcal{A}_0$ which contains the modulation alphabet $\mathcal{A}$ and zero symbol is utilized to denote the active and inactive users. Then the modulated symbol vectors in all n segments are spread by an specific spreading sequence $\mathbf{s}_k^i$ from the tandem spreading combination Ω_k, which is written as

$$\Omega_k = \left\{ \mathbf{s}_k^1, \mathbf{s}_k^2, \mathbf{s}_k^3, \ldots, \mathbf{s}_k^m, \mathbf{s}_k^{m+1} \right\}. \tag{5.2}$$

The tandem spreading combination Ω_k has n sequences with respect to the n segments. The sequences in Ω_k are selected from a primitive orthogonal spreading sequences dictionary $\mathcal{E} = \{ \mathbf{e}_1, \mathbf{e}_2, \ldots, \mathbf{e}_q \}$. The spreading factor of the primitive spreading sequence is q. Additionally, the primitive spreading sequence can be selected repeatedly.

In tandem spreading, a codebook C is designed to contain the combinations for all the served users. For the codebook design, two design criteria are involved. The first criterion is that the codebook size is squared with the spreading factor as

$$\text{card}(C) = q^2. \tag{5.3}$$

With this criterion, the maximum number of served users K_{max} increases quadratically with the spreading factor q. The second criterion confines the maximum number of coincidences to be 1. The coincidence indicates the occurrence that two combinations have the same spreading sequence at the same position, namely $s_k^i = s_l^i$ for $s_k^i \in \Omega_k, s_l^i \in \Omega_l$ and $\forall i \in \{1, 2, \ldots, n\}$. This criterion is represented as $\forall \Omega_k, \Omega_l \in C$

$$Coll(\Omega_k, \Omega_l) \leqslant 1 \tag{5.4}$$

where $Coll(\Omega_k, \Omega_l)$ denotes the coincidence number. The collision between two active users can be represented by those coincidences. Since PLNC coding only generates one redundancy segment, TSNDMA is able to resolve the collision of two active users with this criterion.

Based on the design criteria above, TSNDMA provides an algorithm to generate a qualified tandem spreading codebook. The algorithm is shown in Fig. 5.2.

At first, the dictionary $\mathcal{E}$ is defined to contain q orthogonal spreading sequences. The segment number is set to be equal to the spreading factor as $n = q$. It is easy to find that if the second criterion is satisfied under $n = q$, it is also met under the condition $n < q$. Also, the codebook size is set as $K_{max} = q^2$ according to the first criterion. The codebook C is initialized to be empty. In the algorithm, the i-th spreading sequence element s_k^i of the combination Ω_k is obtained by $s_k^i = e_{\mod (k-1+(i-1)\lceil k/N_c\rceil, N_c)+1}$. The codebook design introduces a proposition shown below.

Proposition *For a prime number q, $\forall k \in [1, K_{max}]$ and $i \in [1, n]$, if the i-th spreading sequence element of combination Ω_k is generated as*

$$s_k^i = e_{\mod (k-1+(i-1)\lceil k/q\rceil, q)+1} \tag{5.5}$$

then $\forall \Omega_k, \Omega_l \in C$.

$$Coll(\Omega_k, \Omega_l) \leqslant 1. \tag{5.6}$$

Furthermore, $\forall k, l \in [1, K_{max}]$, if $\lceil k/q\rceil = \lceil l/q\rceil$, then $Coll(\Omega_k, \Omega_l) = 0$; else if $\lceil k/q\rceil \neq \lceil l/q\rceil$, then $Coll(\Omega_k, \Omega_l) = 1$.   □

The proof of this proposition is written below.

Proof $Coll(\Omega_{k_a}, \Omega_{k_b})$ can be interpreted as the feasible solution number of i_a, i_b to satisfy: $\forall k_a, k_b \in [1, K_{max}]$ and $k_a \neq k_b$, if $\exists i_a, i_b \in [1, n]$ s.t. $s_{k_a}^{i_a} = s_{k_b}^{i_b}$, then $i_a = i_b$. Alternatively, $Coll(\Omega_{k_a}, \Omega_{k_a})$ the solution number of the equation

$$\mod (k_a - 1 + (i - 1)\theta_a, q) = \mod(k_b - 1 + (i - 1)\theta_b, q) \tag{5.7}$$

where $\theta_a = \lceil k_a/q\rceil, \theta_b = \lceil k_b/q\rceil$. Since $k_a, k_b \in [1, K_{max}]$ and $K_{max} = q^2$, $|\theta_a - \theta_b| = |\lceil k_a/N_c\rceil - \lceil k_b/N_c\rceil| < q$. With mathematical manipulations, Eq. (5.7)

Fig. 5.2 Tandem spreading codebook design algorithm

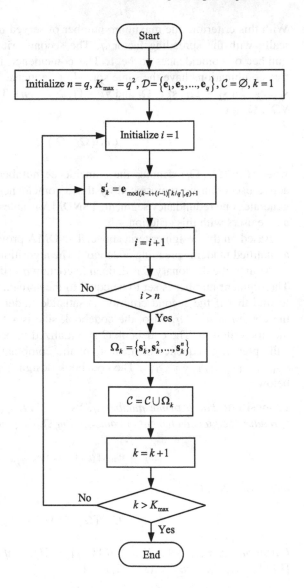

is rewritten as a linear congruence equation as

$$\mod((\theta_b - \theta_a)(i - 1), q) = \mod(k_a - k_b, q). \tag{5.8}$$

According to the theorem in [33], if $\gcd((\theta_b - \theta_a), q) \nmid (k_a - k_b)$, then no solution can be found in Eq. (5.8), and if $\gcd((\theta_b - \theta_a), q) \mid (k_a - k_b)$, then exactly $\gcd((\theta_b - \theta_a), q)$ solutions can be found in Eq. (5.8). Here the notation $\gcd(.)$ denotes the greatest common divisor function. Additionally, q is a prime number, thus if $\theta_a \neq \theta_b$, and $|\theta_a - \theta_b| < q$, then $\gcd((\theta_b - \theta_a), q) = 1$ and $Coll(\Omega_{k_a}, \Omega_{k_b}) = 1$;

else if $\theta_a = \theta_b$, then $|k_a - k_b| < q$ and $\gcd((\theta_b - \theta_a), q) \nmid (k_a - k_b)$ so that $Coll(\Omega_{k_a}, \Omega_{k_b}) = 0$. Therefore, at most one solution is found in the equation and $Coll(\Omega_{k_a}, \Omega_{k_a}) \leqslant 1$. $\qquad \square$

Due to the first criterion, the designed code is specific to the case $K_{\max} = q^2$. Nevertheless, any number of served users should be covered by the design in practice. For an arbitrary K, the spreading factor q is first determined by calculating the minimum prime number which is no smaller than $\sqrt{K}$. Then the codebook C is generated according to the spreading factor q which enables q^2 served users. Hence, for $K \leqslant q^2$, the system only uses the effective codebook $\tilde{C}$ which is composed of the combinations $\{\Omega_1, \Omega_2, \ldots, \Omega_K\}$ in codebook C. At last, the number of segments n should be no larger than q. To achieve this, data packet of each active UE can be divided into sub-packets whose segments is no more than q.

After tandem spreading, the chips of active UE k experience the fading channel. It is assumed that the signals from different active UE arrive synchronously at the receiver. At the BS side, the synthesized signal is first divided into n segments and the i-th segment is written as

$$\mathbf{y}^i = \sum_{k=1}^{K} h_k \mathbf{S}_k^i \mathbf{x}_k^i + \mathbf{n}^i \tag{5.9}$$

where h_k is the complex channel fading for UE k. In this model, slot fading channel is assumed and h_k follows the independent complex Gaussian distribution $\mathcal{CN}(0, 1)$. Moreover, matrix $\mathbf{S}_k^i \in \mathbb{C}^{qb \times b}$ is a diagonal matrix of vector $\mathbf{s}_k^i$ as

$$\mathbf{S}_k^i = \begin{bmatrix} \mathbf{s}_k^i & 0 & \cdots & 0 \\ 0 & \mathbf{s}_k^i & & \vdots \\ \vdots & & \ddots & 0 \\ 0 & \cdots & 0 & \mathbf{s}_k^i \end{bmatrix} \tag{5.10}$$

where $\mathbf{s}_k^i \in \mathbb{C}^{q \times 1}$ is the aforementioned spreading sequence vector. In addition, the vector $\mathbf{n}^i \in \mathbb{C}^{qb \times 1}$ indicates additive white Gaussian noise (AWGN) who follows the distribution $\mathcal{CN}(0, \sigma^2)$ and σ^2 represents the noise power.

According to the codebook design described before, the spreading sequence $\mathbf{s}_k^i$ at any segment $i \in [1, n]$ of user $k \in [1, K]$ is equal to one primitive spreading sequence $\mathbf{e}_j$ in $\mathcal{E}$ for $j \in [1, q]$. Therefore, all those users whose spreading sequences are $\mathbf{e}_j$ at segment i are grouped into a user group $\mathcal{G}_j^i$. In other words, $k \in \mathcal{G}_j^i$ as long as $\mathbf{s}_k^i = \mathbf{e}_j$. Accordingly, all the K served users are partitioned into $\tilde{q}$ user groups where $\tilde{q} \leqslant q$. Based on above, Eq. (5.9) is rewritten as

$$\mathbf{y}^i = \sum_{j=1}^{\tilde{q}} \bar{h}_j \mathbf{\Gamma}_j \mathbf{X}_j^i + \mathbf{n}^i \tag{5.11}$$

where

$$\bar{h}_j = \sum_{k=1}^{K_j^i} h_{\mathcal{G}_j^i(k)} \tag{5.12}$$

$$\mathbf{\Gamma}_j = [\mathbf{E}_j, \mathbf{E}_j, \ldots, \mathbf{E}_j] \tag{5.13}$$

$$\mathbf{E}_j = \begin{bmatrix} \mathbf{e}_j & \mathbf{0} & \cdots & \mathbf{0} \\ \mathbf{0} & \mathbf{e}_j & & \vdots \\ \vdots & & \ddots & \mathbf{0} \\ \mathbf{0} & \cdots & \mathbf{0} & \mathbf{e}_j \end{bmatrix} \tag{5.14}$$

$$\mathbf{X}_j^i = [(\mathbf{x}_{\mathcal{G}_j^i(1)}^i)^T, (\mathbf{x}_{\mathcal{G}_j^i(2)}^i)^T, \ldots, (\mathbf{x}_{\mathcal{G}_j^i(K_j^i)}^i)^T]^T. \tag{5.15}$$

Here $\mathcal{G}_j^i(k)$ denotes the k-th UE in $\mathcal{G}_j^i$ and K_j^i denotes the number of all UEs in group $\mathcal{G}_j^i$ where $\sum_{j=1}^{\tilde{q}} K_j^i = K$ for $i \in [1, n]$.

For the received signal, user identification is implemented through determining the existence of each primitive spreading sequence on each segment. Assume perfect channel states for all the served UEs are known at the receiver, and then the energy detection is implemented as

$$r_j^i = \sum (\bar{h}_j \mathbf{\Gamma}_j)^H \mathbf{y}^i \tag{5.16}$$

where r_j^i is the correlation between $\mathbf{e}_j$ and the received signal at segment i. The user identification algorithm is shown in Fig. 5.3.

In this algorithm, a threshold Th is set to detect the existence of the primitive spreading sequence at each segment from the received signal. If $r_j^i > Th$, then it can be determined that the a least one UE in $\mathcal{G}_j^i$ is involved in the received signal. Accordingly, the algorithm has $\mathcal{U}^i = \mathcal{U}^i \cup \mathcal{G}_j^i$ to count for the potential active UEs for segment i. At last, the intersection $\mathcal{U}$ of all the $\mathcal{U}^i$ for $i \in [1, T_s]$ is obtained. Then we have a proposition for this algorithm shown below.

Proposition *Assume energy detection is perfect for each primitive spreading sequence at each segment, if the active UE number is less than the segment number, namely $K_a < n$, then the obtained active UE set $\mathcal{U}$ is the actual active UE set $\mathcal{U}_a$, namely $\mathcal{U} = \mathcal{U}_a$.* □

Fig. 5.3 User identification algorithm

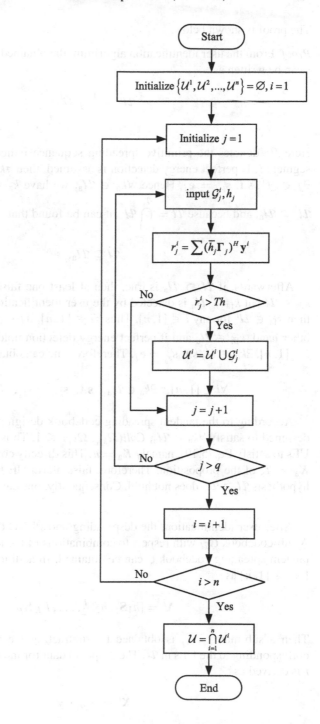

The proof is shown below.

Proof From the user identification algorithm, the obtained active UE set at segment i can be written as

$$\mathcal{U}^i = \bigcup_{j \in \mathcal{J}^i} \mathcal{G}^i_j. \tag{5.17}$$

Here $\mathcal{J}^i$ denotes the primitive spreading sequence indices which are detected at segment i. If perfect energy detection is assumed, then $\forall k_a \in \mathcal{U}_a$, and $\forall i \in [1, n]$, $\exists j \in \mathcal{J}^i$, s.t. $\mathbf{s}^i_{k_a} = \mathbf{e}_j$. Hence, $\forall k_a \in \mathcal{U}_a$, we have $k_a \in \mathcal{U}^i$. Thus $\forall i \in [1, n]$, $\mathcal{U}^i \supseteq \mathcal{U}_a$, and because $\mathcal{U} = \bigcap_{i=1}^{T_s} \mathcal{U}^i$, it can be found that

$$\mathcal{U} \supseteq \mathcal{U}_a. \tag{5.18}$$

Afterwards, if $\mathcal{U} \supset \mathcal{U}_a$ is true, then at least one false alarm UE k_{fa} such that $k_{fa} \in \mathcal{U}$ but $k_{fa} \notin \mathcal{U}_a$ is detected by the user identification algorithm. If $k_{fa} \in \mathcal{U}$, then $k_{fa} \in \mathcal{U}^i$ for any $i \in [1, n]$. Thus $\forall i \in [1, n]$, $\exists j \in \mathcal{J}^i$, s.t. $\mathbf{s}^i_{k_{fa}} = \mathbf{e}_j$. On the other hand, $k_{fa} \notin \mathcal{U}_a$ and if perfect energy detection holds, then $\forall j \in \mathcal{J}^i$ for any $i \in [1, n]$, $\exists k_a \in \mathcal{U}_a$ s.t. $\mathbf{s}^i_{k_a} = \mathbf{e}_j$. Therefore, one can obtain that

$$\forall i \in [1, n], \quad \exists k_a \in \mathcal{U}_a, \quad \text{s.t.} \quad \mathbf{s}^i_{k_{fa}} = \mathbf{s}^i_{k_a}, \quad k_{fa} \neq k_a. \tag{5.19}$$

According to the tandem spreading codebook design criterion, the codebook is designed to satisfy $\forall k_a \in \mathcal{U}_a$, $Coll(\Omega_{k_{fa}}, \Omega_{k_a}) \leqslant 1$. Thus $\mathcal{U}_a$ should have n active UEs to satisfy Eq. (5.19), namely $K_a = n$. This directly contradicts the precondition $K_a < T_s$ of the proposition. Therefore, false alarm UE k_{fa} does not exist and the hypothesis $\mathcal{U} \supset \mathcal{U}_a$ does not hold. Consequently, one can conclude that $\mathcal{U} = \mathcal{U}_a$

$$\square$$

After user identification, the despreading and PLNC decoding is implemented. A sub-codebook $C_\mathcal{U}$ with respect to combinations of the identified UE set $\mathcal{U}$ in the tandem spreading codebook C can be obtained. In addition, a matrix $\mathbf{A}^i$ is defined for $i \in [1, n]$ as

$$\mathbf{A}^i = [h_1 \mathbf{S}^i_1, h_2 \mathbf{S}^i_2, \ldots, h_K \mathbf{S}^i_K]. \tag{5.20}$$

Then a sub-matrix $\mathbf{A}^i_\mathcal{U}$ is obtained by subtracting the columns in $\mathbf{A}$ which are corresponding to the UEs in $\mathcal{U}$. The despread data for the identified UEs at segment i is derived as

$$\hat{\mathbf{X}}^i = (\mathbf{A}^i_\mathcal{U})^\dagger \mathbf{y}^i \tag{5.21}$$

where $\hat{\mathbf{X}}^i = [(\mathbf{x}^i_{\mathcal{U}(1)})^T, (\mathbf{x}^i_{\mathcal{U}(2)})^T, \ldots, (\mathbf{x}^i_{\mathcal{U}(K_u)})^T]^T$ and (K_u) is the number of identified UEs.

After despreading, the despread data $\hat{\mathbf{x}}^i_{\mathcal{U}(k)}$ of the k-th UE in $\mathcal{U}$ is demodulated to get the bits as $\hat{\mathbf{d}}^i_{\mathcal{U}(k)}$ for segment $i \in [1, n]$. Then PLNC decoding is applied to recover the colliding segment. In this decoding, first the colliding segments of each identified UEs are determined according to the sub-codebook $C_{\mathcal{U}}$. Then the receiver finds the decodable UEs who have no more than 2 colliding segments and abandon the unqualified UEs. At last, for each decodable UE, the bits in the colliding segment are recovered by modulo-2 summing of the decoded bits at other segments to finish the PLNC decoding.

In the rest of this section, the user identification and data detection performance are evaluated through numerical simulations. In the setup, the served UE number is ranging from 5 to 25 and the active UE number K_a is 2 and 3. Additionally, transmission SNR is set to be 20 dB to mitigate the influence of noise in energy detection. The spreading factor q here is 5. In the meantime, the segment number in TSNDMA is set to be 5 as well. First, the false alarm probability of user identification in TSNDMA and CSMUD are shown in Figs. 5.4 and 5.5 respectively. For the false alarm performance of TSNDMA, one can clearly observe that probabilities for all the conditions are extremely low except the case when $K_a = 5$ and $K > q$. Thus this simulation result verifies the user identification proposition. For CSMUD, except the case of $K = q$, the probabilities are relatively high. It is because the spreading sequences in this comparison are short and the sparsity of user activity is limited so that multiple access interference (MAI) cannot be well mitigated. Note that the false alarm probability for CSMUD is still increasing with K for the case $K_a = 1$. Due to the cross correlation between the non-orthogonal spreading sequences in CSMUD, the receiver is possible to determine an inactive user as active. Hence, it can be considered that when $K_a = 1$, MAI from the inactive

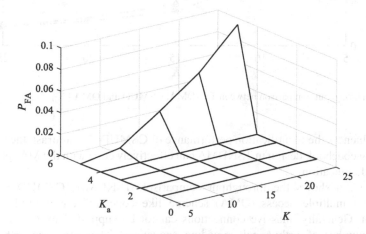

Fig. 5.4 False alarm probability of the user identification in TSNDMA [30]

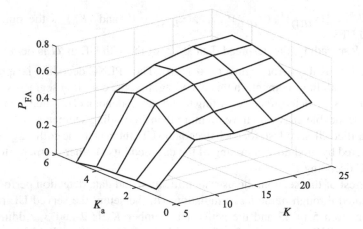

Fig. 5.5 False alarm probability of the user identification in CS-MUD [30]

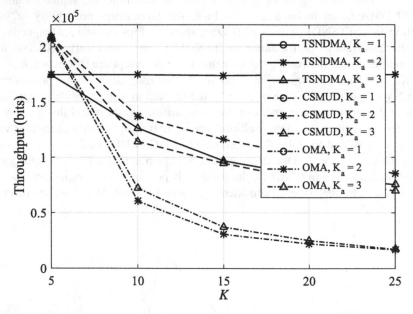

Fig. 5.6 Throughput comparison between TSNDMA, CSMUD and OMA [30]

users influences the identification performance of CSMUD. By contrast, the tandem spreading codebook design in TSNDMA can effectively cope with the MAI problem in user identification.

Figure 5.6 shows the throughput comparison TSNDMA, CSMUD and the orthogonal multiple access (OMA) scheme like code division multiple access (CDMA). Generally, massive connections cannot be supported by OMA due to limited number of orthogonal spreading sequences. Therefore, the orthogonal spreading sequences in OMA are allocated repeatedly to the excessive served UEs. In the comparison, different Ks and K_as are considered. In the simulation, the ideal throughput for each active UE is 2.2×10^5 for CSMUD and OMA. On the other hand, because PLNC redundancy segment is added, the ideal throughput for each active UE in TSNDMA is 1.76×10^5. The first observation is that when $K = 5$, the per active UE throughputs of TSNDMA, CSMUD and OMA are asymptotically ideal. In this case, the served UE number is equal to the spreading factor so that MAI is absence to degrade the detection performance. Also, due to the PLNC redundancy, TSNDMA has a lower throughput than CSMUD and OMA in this case. Nonetheless, while the served UEs is larger than the spreading factor, the MAI exists to influence the detection performance of CSMUD and OMA. The influence is increasing with K and K_a as shown in the figure. For TSNDMA, it can be surprisingly observed that the throughput remains unchanged with the increasing K under the case $K_a = 2$. It is because that two active UEs have at most one colliding segment, which can be recovered by the PLNC redundancy segment. For the case $K_a = 3$ in TSNDMA, the per active UE throughput is decreasing with K. For three active UEs, two colliding segments are possibly generated. Single PLNC redundancy segment cannot recover two colliding segments so that throughput loss appears due to MAI.

Finally, the evaluations with different active probability p_a under the conditions of $K = 5$, 10 and 15 are shown by the plots in Fig. 5.7. For $K = 5$, the spreading sequences are orthogonal, thus a similar result can be obtained that the throughput of TSNDMA is inferior to those of CSMUD and OMA. While $K = 10$ or 15, one can observe that TSNDMA can basically keep the advantage on both CSMUD and OMA. One exception is the case $K = 15$ and $p_a > 0.25$, the throughput of TSNDMA falls below that of CSMUD. In this case, the active UE number becomes large and thereby the collisions cannot be solved by single PLNC redundancy.

5.2 Coded Tandem Spreading Multiple Access (CTSMA)

CTSMA is a generalized version of TSNDMA. There have two differences between CTSMA and TSNDMA. The first is that multiple redundancy segments are enabled in CTSMA to support more active UEs. The second is that a more flexible tandem spreading codebook design is applied in CTSMA.

Figure 5.8 provides the system structure of CTSMA, which is similar to the structure of TSNDMA except the PLNC coding is substituted by the segment coding. At the transmitter side, the b channel coded data bits of UE k are divided into m segments as $\mathbf{D}_{k,m} = \{\mathbf{d}_{k,1}, \mathbf{d}_{k,2}, \ldots, \mathbf{d}_{k,m}\}$ under the assumption that b can be divided by m. Then the segment coding is implemented to encode those m segments into n coded segments as $\mathbf{D}_{k,n} = \{\mathbf{d}_{k,1}, \mathbf{d}_{k,2}, \ldots, \mathbf{d}_{k,n}\}$. In segment coding, Reed-Solomon (RS) code is applied. RS code is a maximum distance separable (MDS) erasure code for any values of m and n. It means that segment coding is able to recover any $R = n - m$ colliding segments [34]. In realizing the RS encoding, each

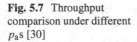

Fig. 5.7 Throughput comparison under different p_as [30]

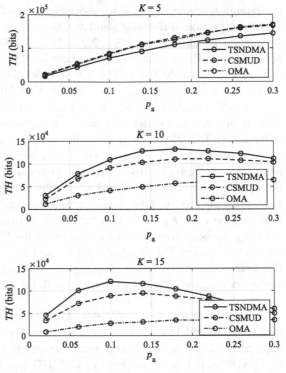

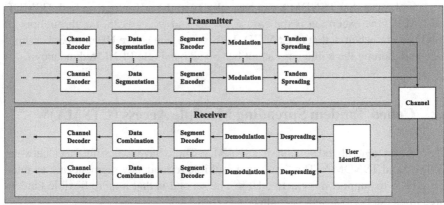

Fig. 5.8 System structure of CTSMA [31]

segment in $\mathbf{D}_{k,m}$ is mapped to the Galois field $\text{GF}\left(2^{\frac{b}{m}}\right)$. Then $\mathbf{D}_{k,m}$ is converted to $\mathbf{\Gamma}_{k,m}$ as

$$\mathbf{\Gamma}_{k,n} = \mathbf{\Gamma}_{k,m}\mathbf{G} \quad \text{in } \text{GF}\left(2^{\frac{b}{m}}\right) \tag{5.22}$$

Fig. 5.9 An illustrative example of segment coding in CTSMA [31]

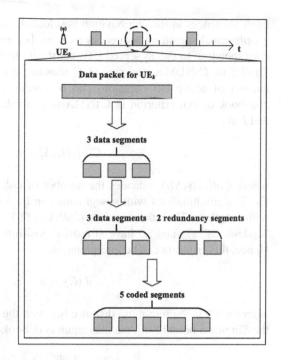

where $\mathbf{G} \in \mathrm{GF}\left(2^{b_m}\right)^{m \times n}$ is the generator matrix and the n coded segments are represented by $\mathbf{\Gamma}_{k,n}$ in Galois field. Here Vandermonde matrix is utilized as the generator matrix [35]. The generator is written below.

$$\mathbf{G} = \begin{pmatrix} \alpha_0^0 & \alpha_1^0 & \cdots & \alpha_{n-2}^0 & \alpha_{n-1}^0 \\ \alpha_0^1 & \alpha_1^1 & \cdots & \alpha_{n-2}^1 & \alpha_{n-1}^1 \\ \alpha_0^2 & \alpha_1^2 & \cdots & \alpha_{n-2}^2 & \alpha_{n-1}^2 \\ \vdots & \vdots & \ddots & \vdots & \vdots \\ \alpha_0^{m-1} & \alpha_1^{m-1} & \cdots & \alpha_{n-2}^{m-1} & \alpha_{n-1}^{m-1} \end{pmatrix}. \tag{5.23}$$

The elements $\{\alpha_0, \alpha_1, \ldots, \alpha_{n-1}\}$ in $\mathbf{G}$ should be distinct in $\mathrm{GF}\left(2^{\frac{b}{m}}\right)$ and thereby $n \leqslant 2^{\frac{b}{m}}$. Afterwards, the binary bits are obtained by converting $\mathbf{\Gamma}_{k,n}$ back to the binary form. An example with parameters $m = 3$ and $n = 5$ is shown in Fig. 5.9. In this example, erasure code is applied so that the linear combinations of the three data segments yield the two redundancy segments.

After segment encoding, n segments data symbols $\{\mathbf{x}_{k,1}, \mathbf{x}_{k,2}, \ldots, \mathbf{x}_{k,n}\}$ are obtained from the n coded segments data bits through modulation. Subsequently, multiple spreading sequences are employed to spread the data symbols at different segments in tandem spreading. Those spreading sequences for all the segments of UE k form the tandem spreading combination as $\Omega_k = \{\mathbf{s}_{k,1}, \mathbf{s}_{k,2}, \ldots, \mathbf{s}_{k,n}\}$.

Each sequences in the combination is selected from the dictionary which contains q orthogonal spreading sequences as $\mathcal{E} = \{e_1, e_2, \ldots, e_q\}$. All the combinations corresponding to the K served UEs constitute to a tandem spreading codebook C. Similar to TSNDMA, the size of C should be maximized on condition that the number of active UE collisions can be confined. Different with TSNDMA, the codebook design criterion in CTSMA is generalized, which is that $\forall \Omega_k, \Omega_l \in C$ and $k \neq l$

$$Coll(\Omega_k, \Omega_l) \leqslant r \tag{5.24}$$

where $Coll(\Omega_k, \Omega_l)$ indicates the number of colliding segments between Ω_k and Ω_l. The combinations with n segments can be analogous to the q-ary codewords with length n. Then the criterion $Coll(\Omega_k, \Omega_l) \leqslant r$ can be represented that the "codewords" Ω_k and Ω_l have at most r positions occupied by the same element. Hence, this criterion can be interpreted as

$$d(C) = n - r. \tag{5.25}$$

where $d(C)$ is the minimum distance between the "codewords" in C. According to the Singleton bound [36], the maximum codebook size is written as

$$|C| \leqslant q^{n-d(C)+1} = q^{r+1}. \tag{5.26}$$

If the codewords in C are MDS codes, the equality holds [35]. Accordingly, C can be generated through the q-ary $(n, r+1, n-r)$ RS code with a prime power q. Similarly, the RS code generator matrix is a Vandermonde matrix $M \in GF(q)^{(r+1)\times n}$ shown as

$$M = \begin{pmatrix} \beta_0^0 & \beta_1^0 & \cdots & \beta_{n-2}^0 & \beta_{n-1}^0 \\ \beta_0^1 & \beta_1^1 & \cdots & \beta_{n-2}^1 & \beta_{n-1}^1 \\ \beta_0^2 & \beta_1^2 & \cdots & \beta_{n-2}^2 & \beta_{n-1}^2 \\ \vdots & \vdots & \ddots & \vdots & \vdots \\ \beta_0^r & \beta_1^r & \cdots & \beta_{n-2}^r & \beta_{n-1}^r \end{pmatrix}. \tag{5.27}$$

Also, the elements $\{\beta_0, \beta_1, \ldots, \beta_{n-1}\}$ are distinct in $GF(q)$ so that $n \leqslant q$. In the meantime, all the possible combinations with $r+1$ elements in $GF(q)$ compose an elementary codebook C_0. Then the construction of C is shown as

$$C = C_0 M \quad \text{in } GF(q). \tag{5.28}$$

Here the rows in the matrices C and C_0 represent the combinations in C and C_0 respectively. In addition, since C contains q^{r+1} tandem spreading combinations, the served UE number $K \leqslant q^{r+1}$. If $K = q^{r+1}$, then all the combinations in C are allocated to the UEs. Otherwise, the UE are allocated by a subset of combinations in

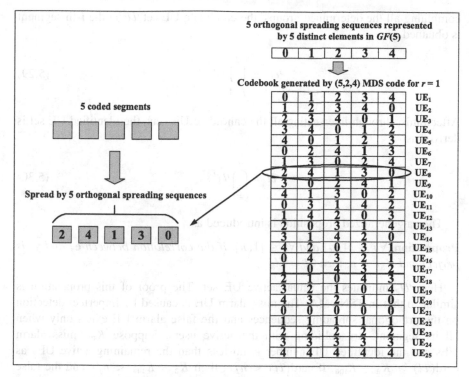

5 coded segments

Spread by 5 orthogonal spreading sequences

5 orthogonal spreading sequences represented by 5 distinct elements in $GF(5)$

0	1	2	3	4

Codebook generated by (5,2,4) MDS code for $r = 1$

0	1	2	3	4	
0	1	2	3	4	UE₁
1	2	3	4	0	UE₂
2	3	4	0	1	UE₃
3	4	0	1	2	UE₄
4	0	1	2	3	UE₅
0	2	4	1	3	UE₆
1	3	0	2	4	UE₇
2	4	1	3	0	UE₈
3	0	2	4	1	UE₉
4	1	3	0	2	UE₁₀
0	3	1	4	2	UE₁₁
1	4	2	0	3	UE₁₂
2	0	3	1	4	UE₁₃
3	1	4	2	0	UE₁₄
4	2	0	3	1	UE₁₅
0	4	3	2	1	UE₁₆
1	0	4	3	2	UE₁₇
2	1	0	4	3	UE₁₈
3	2	1	0	4	UE₁₉
4	3	2	1	0	UE₂₀
0	0	0	0	0	UE₂₁
1	1	1	1	1	UE₂₂
2	2	2	2	2	UE₂₃
3	3	3	3	3	UE₂₄
4	4	4	4	4	UE₂₅

2	4	1	3	0

Fig. 5.10 An illustrative example of tandem spreading codebook design in CTSMA [31]

C. Hence, the parameter r is set as $r = \lceil \log_q K - 1 \rceil$, where $\lceil . \rceil$ denotes the ceiling function. Figure 5.10 shows an example with parameters $q = 5$, $n = 5$ and $r = 1$. The elements in GF(5) represents the 5 orthogonal spreading sequences. A (5,2,4) RS code is utilized to construct the set C. In this example, 25 qualified combinations are contained in this codebook so that all the 25 UEs can be allocated. If UE k is active, its five coded segments are tandemly spread the allocated combination.

Originally, MDS code is employed for error detection and correction. In the tandem spreading codebook, the design criterion agrees with the feature of minimum Hamming distance and MDS code can achieve the maximum code size for a given minimum Hamming distance. Therefore, MDS code is applied to maximize the tandem spreading codebook size in order to support a large number of served UEs.

At the receiver side, n segments are obtained by dividing the received signal for the following user identification and segment decoding. In user identification, each segment of the received signal is correlated with all the orthogonal spreading sequences in $\mathcal{E}$. The tandem spreading codebook C is pre-defined at BS and UEs in CTSMA. The UE group $\mathcal{G}_j^i$ for $j \in [1, q]$ is obtained by collecting the served UEs whose assigned spreading sequence is $\mathbf{e}_j$ at segment i. If $\mathbf{e}_j$ is detected at segment i of the received signal, it can be determined that at least one UE in $\mathcal{G}_j^i$ are active. By combining all the determined groups, the candidate UE set $\mathcal{U}^i$ at the i-th segment is obtained as

$$\mathcal{U}^i = \bigcup_{j=1}^{q} \mathcal{G}_j^i. \tag{5.29}$$

Afterwards, through intersecting all the candidate UE sets, the identified UE set is derived as

$$\mathcal{U} = \bigcap_{i=1}^{n} \mathcal{U}^i. \tag{5.30}$$

Then a generalized proposition is introduced as

Proposition $\forall j \in [1, q]$ *and* $\forall i \in [1, n]$, *if the correlation between* $\mathbf{e}_j$ *and* $\mathbf{y}_i$ *is perfect and* $K_a r < n$, *then* $\mathcal{U} = \mathcal{U}_a$

Here $\mathcal{U}_a$ indicates the actual active UE set. The proof of this proposition is similar to that in TSNDMA. The miss alarm UE is caused by imperfect detection on the orthogonal spreading sequences and the false alarm UE exists only when all its segments are colliding with the active users. Suppose K_{ma} miss alarm UEs and the identified UE number is no less than the remaining active UEs as card($\mathcal{U}$) $\geqslant K_a - K_{ma}$. If card($\mathcal{U}$) $< n/r$, then $K_a - K_{ma} < n/r$ and the false alarm UE whose all segments are colliding with the remaining active users cannot be found. Therefore, a corollary can be obtained as

Corollary *If the identified UE number card($\mathcal{U}$) $< n/r$, then the false alarm UE does not exist.*   □

An illustrative example for user identification is shown in Fig. 5.11. **UE$_8$** where **UE$_{12}$**, **UE$_{20}$** are active. If perfect energy detection is assumed, the detected spreading sequences are applied to obtain the candidate UE group for each segment. At last, those candidate UE groups for all 5 segments are intersected to identify **UE$_8$**, **UE$_{12}$**, **UE$_{20}$**, which are actually the active UEs.

After user identification, the receiver despreads the received signal of identified UE k with the sequences from the corresponding combination in C. After demodulation, $n - m$ segments which includes all the colliding segments are deleted to form the reduced binary form segments. Similar to the segment encoding, the segments in $\bar{\mathbf{D}}_{k,m}$ are converted to the GF$\left(2^{\frac{b}{m}}\right)$ elements as $\bar{\mathbf{\Gamma}}_{k,m}$, which is realized as

$$\hat{\mathbf{\Gamma}}_{k,m} = \bar{\mathbf{\Gamma}}_{k,m} \bar{\mathbf{G}}^{-1} \quad \text{in GF}\left(2^{\frac{b}{m}}\right). \tag{5.31}$$

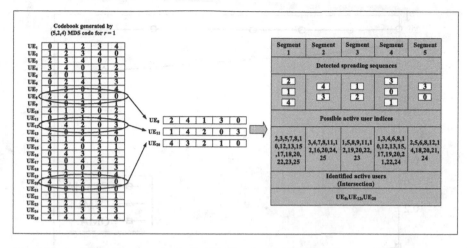

Fig. 5.11 An illustrative example of user identification in CTSMA [31]

Here $\bar{\mathbf{G}} \in \mathrm{GF}\left(2^{\frac{b}{m}}\right)^{m \times m}$ denotes a sub-matrix of $\mathbf{G}$ where the $n - m$ columns corresponding to the deleted segments are deleted. Afterwards, $\hat{\mathbf{D}}_{k,m}$ is obtained by converting back $\hat{\mathbf{\Gamma}}_{k,m}$. At last, the receiver combines all the m decoded segments $\hat{\mathbf{D}}_{k,m}$ to form a data packet for the following channel decoding.

Based on the description of CTSMA shown above, the influence of collision is controlled by the tandem spreading codebook and resolved by segment coding. For K_a active UEs, each UE has maximum $(K_a - 1)r$ colliding segments. In order to resolve the collision, the number of redundancy segments should satisfy

$$n - m \geqslant (K_a - 1)r. \tag{5.32}$$

According to the user identification proposition, as long as perfect energy detection is assumed and $K_a r < n$, K_a active UEs can be identified. In addition, n is confined as $n \leqslant q$, combining with Eq. (5.32), n can be set as $n = q$ to maximize K_a. Also, r is determined as $r = \lceil \log_q K - 1 \rceil$. Hence, for $K > q$, the collision can be resolved if

$$K_a \leqslant \min\left(\frac{q(1-\rho)}{\lceil \log_q K - 1 \rceil} + 1, \frac{q-1}{\lceil \log_q K - 1 \rceil}\right). \tag{5.33}$$

Here $\rho = \frac{m}{n}$ is the segment coding rate and $1 - \rho$ represents the UE data rate loss. In TSNDMA, the code rate is $\rho_{\text{TSNDMA}} = \frac{m}{m+1}$ and the rate loss can be found in the throughput comparison as shown in Fig. 5.6.

Numerical simulations are conducted to evaluate the performance of CTSMA. In order to represent massive connections, the served UE number K to spreading factor

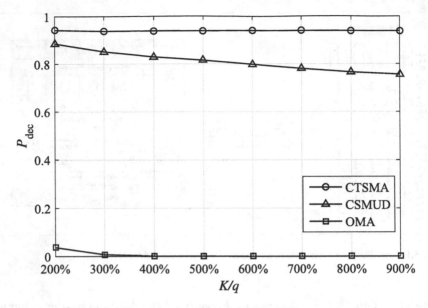

Fig. 5.12 Decoding probability comparison for 4 active UEs [13]

q ratio is introduced as K/q. Here the spreading factor q represents the number of available radio resources.

First, the capability of a deterministic number of active UEs under different K/qs is evaluated. The active UE number is fixed in each time slot in this evaluation. The resource number is set as $q = 9$ and the size of information is set as 27. Moreover, channel coding applies the convolutional code whose code rate is 0.75 so that the size of channel coded bits is $b = 36$. In segment coding, the message segment size is set as $m = 6$ and the coded segments is set to have 9 segments, which is equal to the spreading factor, namely $n = q = 9$. Therefore, the overall coding rate includes channel coding and segment coding in CTSMA is 0.5. In this evaluation, CSMUD and OMA are involved as the benchmark. Since CSMUD and OMA do not apply segment coding, the channel coding is convolutional coding with rate 0.5. In the meanwhile, BPSK modulation and Rayleigh flat fading channel are assumed.

Figure 5.12 shows the comparison among CTSMA, CSMUD and OMA on collision resolution probability for 4 active UEs in each time slot. The collision resolution probability P_{dec} is defined as the probability that all the 4 active UEs in a slot are successfully decoded. In OMA, unrecoverable collisions are very likely to occur because of orthogonal resource allocation. It can be concluded that conventional OMA scheme is not suitable for the grant-free random access system with a large number of served UEs. Afterwards, in CSMUD, the collision resolution probability is falling with K/q. Although the UE activity of a fixed number of active UEs is more sparse at the high K/q region, the performance is still limited by the MAI from the cross-correlation of non-orthogonal spreading sequences. In CTSMA,

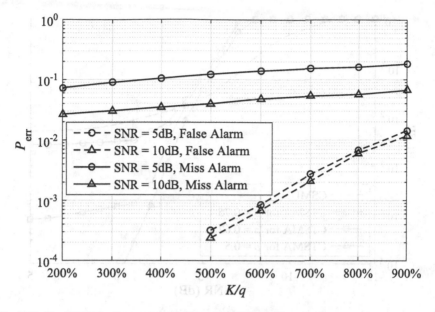

Fig. 5.13 User identification error probability of CTSMA [13]

the system is able to effectively control and resolve the colliding segments with the tandem spreading codebook and segment coding. Hence, the performance of CTSMA for a fixed active UE number keeps unchanged for $K/q \in [200\%, 900\%]$ and we can conclude that CTSMA has the robustness to the collision under a wide range of connectivity.

Subsequently, stochastic active UE number with active probability $p_a = 0.05$ is considered to evaluate the user identification performance of CTSMA. The false alarm and miss alarm probabilities of CTSMA are shown in Fig. 5.13. The false alarm probability increases with K/q but varies slightly with SNR. The false alarm probability mainly depends on the active UE number. In this evaluation, the number of maximum colliding segments between any two users is $r = 1$ for $K/q \in [200\%, 900\%]$. Thus the false alarm probability is mainly determined by the probability that $K_a < 9$. On the other side, since miss alarm is mainly caused by the imperfect energy detection, the MA probability is sensitive to SNR rather than K/q. In this figure, the miss alarm due to Rayleigh fading is the dominant factor to influence the user identification performance.

In the following figures, the evaluation is conducted under the AWGN channel. Also, the served UE number and the active probability are set as $K = 64$ and $p_a = 0.02$ respectively. The spreading factor is $q = 16$ and the overall code rate is set as $\frac{1}{2}$ Here ρ is the segment coding rate, convolutional code with rates $\frac{1}{2\rho}$ is applied in CTSMA.

Figure 5.14 shows the BLER performance of CTSMA with different ρs. The first observation is that at the low SNR region, CSMUD outperforms CTSMA. In

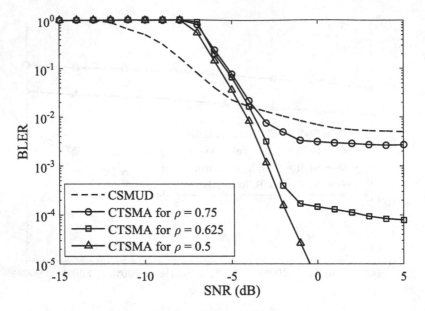

Fig. 5.14 Performance comparison between CTSMA and CSMUD [31]

this region, the detection errors due to noise are dominant and CTSMA cannot have the same channel coding gain as CSMUD due to the lower channel coding rate. In addition, the number of samples for energy detection is reduced by data segmentation in CTSMA to degrade the user identification performance. However, at high SNR region, collision is dominant to influence the system performance so that it can be observed that CTSMA outperforms CSMUD. If the instantaneous K_a satisfies inequality (5.33), the influence of collision in CTSMA can be eliminated. The error floor of CTSMA is determined by the probability the inequality (5.33) is met. Therefore, CTSMA can achieve high transmission reliability among the massive served UEs by manipulating the segment coding rate ρ.

Figure 5.15 illustrates the tradeoff among the served UE number, the collision resolution probability and the UE data rate in CTSMA. They represent the connectivity, reliability and data rate respectively. Here the collision resolution probability is denoted by P. It is the binomial cumulative distribution function (CDF) of K_a on condition that $K_a > 0$. With low data rate, both high reliability and connectivity can be guaranteed. It is noted that the available number of served UEs on each reliability level is upper bounded at the low segment coding rate region, which is caused by user identification as shown in inequality (5.33).

Based on the analysis, CTSMA is an approach to accomplish both the requirements of connectivity and reliability if low UE data rate is allowed. CTSMA has the weakness on energy detection in the low SNR region. However, high SNR has to be present to realize reliable communications. Accordingly, CTSMA is suitable for

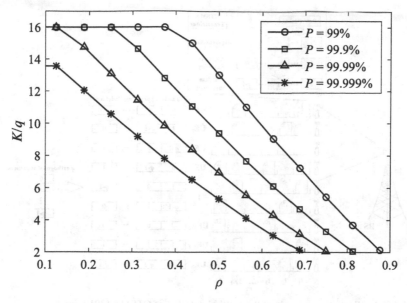

Fig. 5.15 The tradeoff among the served UE number, the collision resolution probability and the UE data rate in CTSMA [31]

the mMTC applications who demand high reliability rather than an extremely low transmission power.

5.3 Multi-slot Design of CTSMA

This section introduces a multi-slot design scheme of CTSMA. In order to achieve a higher collision resolution capability, this scheme combines CTSMA and CSA. An illustrative example of the multi-slot design is shown in Fig. 5.16. According to aforementioned descriptions, each UE in CSA transmits the data replicas by the randomly selected slots in the uplink contention period. CTSMA transmitter processes the replica through segmentation, segment encoding and tandem spreading. In the figure, the replica which is denoted as $\mathbf{D}_k(t_\tau)$ at slot t_τ for $\tau \in [1, T]$ is divided into m data segments. Then the data segments are encoded to yield $n - m$ redundancy segment. Meanwhile, the tandem spreading codebook is generated by employing q orthogonal spreading sequences. With parameter r, q^{r+1} combinations are contained so that the codebook can serve K UEs.

At the BS side, the signals from all the uplink contention period slots are collected by the receiver. Subsequently, inter-slot SIC is employed to resolve the collision. According to Eq. (5.33) in CTSMA, if identified UE number in one slot is no more than K_d as

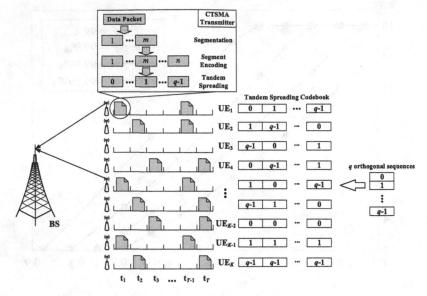

Fig. 5.16 An illustrative example of the multi-slot design of CTSMA [32]

$$K_{\mathrm{d}} = \min\left(\frac{q(1-\rho)}{\lceil \log_q K - 1 \rceil} + 1, \frac{q-1}{\lceil \log_q K - 1 \rceil} \right) \qquad (5.34)$$

the false alarm UE does not exist and this slot can be considered as a collision-free slot, namely singleton slot in CSA. For the instantaneous active UE number K_{a} on condition that $K_{\mathrm{a}} \geqslant 1$, the probability of a collision-free slot can be written as an indicator function as

$$\mathcal{I}\left(K_{\mathrm{a}} \mid K_{\mathrm{a}} \geqslant 1\right) = \begin{cases} 1, & K_{\mathrm{a}} \leqslant K_{\mathrm{d}} \\ 0, & K_{\mathrm{a}} > K_{\mathrm{d}} \end{cases}. \qquad (5.35)$$

For one active UE, the probability that one of its occupied slots has K_{a} active UEs is represented as $\Omega_{K_{\mathrm{a}}}$, which is shown in Eq. (4.8). Then one can denote the average UE collision resolution probability for one slot as

$$\bar{P}_{\mathrm{CTSMA}} = \sum_{K_{\mathrm{a}}=1}^{K_{\mathrm{a,dec}}} \Omega_{K_{\mathrm{a}}}. \qquad (5.36)$$

Based on the statements above, the user identification and K_{d} in CTSMA can be utilized to find the singleton slot for data detection and deletes all the replicas of the decoded UEs in each SIC iteration. The corresponding receiver process is shown in Fig. 5.17.

Fig. 5.17 Collision
resolution algorithm of the
multi-slot design

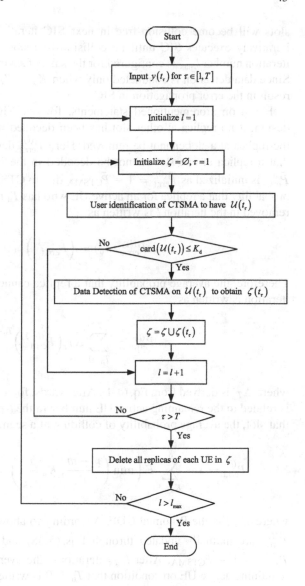

In this algorithm, the inputs are the received signals $y(t_\tau)$ for all the slots $\tau \in [1, T]$. In SIC iteration $l \in [1, l_{max}]$, first the identified UEs $\mathcal{U}(t_\tau)$ at each slot t_τ are obtained by adopting the user identification in CTSMA. Then the collision condition of slot t_τ is determined by observing K_d derived from Eq. (5.34). If the correspond active UE number $K_a \leqslant K_d$, then this slot can be considered as collision-free and thereby the decoded UEs $\zeta(t_\tau)$ are obtained via data detection. The whole decoded UE set ζ for T slots in each SIC iteration is derived as $\zeta = \bigcup_{\tau=1}^{T} \zeta(t_\tau)$. As long as each UE in $\mathcal{U}$ is decoded, the slot positions of other replicas of this UE are known by the receiver. Thus all the replicas of each UE in ζ are deleted to finish one SIC iteration. After removing those replicas of the decoded UEs, some of the colliding slots will become collision-free in next SIC iteration. To this end, the receiver iteratively executes SIC until no collision-free slot is yielded. A maximum SIC iteration number l_{max} is configured for the sake of saving computational complexity. Since data detection is conducted only when $K_a \leqslant K_d$, false alarm UE is absent to result in the error propagation in SIC.

Based on aforementioned statements, for one SIC iteration of the multi-slot design, if no replica at other slot has been decoded in the previous SIC iteration, the replica at a slot cannot be removed. Here $\bar{P}_{err,t}^{(l)}$ denotes the average probability that a replica at one slot cannot be decoded in the l-th SIC iteration for $l \geqslant 1$. $\bar{P}_{err,t}^{(l)}$ is initialized as $\bar{P}_{err,t}^{(0)} = 1 - \bar{P}_{CTSMA}$ due to CTSMA. Similar to Eq. (4.2), the probability that a replica for an active UE who has T_u replicas on one slot cannot be removed in the iteration l is written as

$$P_{err,u}^{(l)}(T_u) = \left(\bar{P}_{err,t}^{(l-1)}\right)^{T_u-1}. \tag{5.37}$$

Therefore, the average probability that a replica cannot be removed in the l-th SIC iteration is written as

$$\bar{P}_{err,u}^{(l)} = \sum_{T_u=1}^{T} \Lambda_{T_u}\left(\bar{P}_{err,t}^{(l-1)}\right)^{T_u-1} \tag{5.38}$$

where Λ_{T_u} is derived from Eq. (4.4). Afterwards, for a slot with K_a active UEs, $\bar{P}_{err,t}^{(l)}$ is related to the remaining active UE number in the l-th iteration. For one replica at that slot, the average probability of colliding at a segment is shown as

$$\bar{P}_{err,t}^{(l)} = 1 - \sum_{K_a=1}^{K} F\left(\min\left(\frac{n-m}{r}, K_a - 1\right); K_a - 1, \bar{P}_{err,u}^{(l)}\right) \Omega_{K_a} \tag{5.39}$$

where $F(.)$ is the Binomial CDF. According to above, one can see that $\bar{P}_{err,t}^{(l)}$ and $\bar{P}_{err,u}^{(l)}$ are mutually updated through Eqs. (5.38) and (5.39). The initialization is $\bar{P}_{err,t}^{(0)} = 1 - \bar{P}_{CTSMA}$. After l_{max} iterations, the average error probability of one remaining active UE on condition that $T_u > 0$ is written as

$$\bar{P}_e = \frac{\sum_{T_u=1}^{T} \Pr(T_u; T, p_a)\left(\bar{P}_{err,t}^{(l_{max})}\right)^{T_u}}{1 - F(0; T, p_a)} \tag{5.40}$$

where $\Pr(.)$ is the Binomial PDF. Therefore, in this multi-slot design, the average UE collision resolution probability $\bar{P}_{MS}$ is represented as

$$\bar{P}_{MS} = 1 - \bar{P}_e. \tag{5.41}$$

Then the average number of UEs resolved from the collision is $\bar{P}_{MS}Kp_a$. The corresponding throughput is defined as

$$TH = \frac{m\bar{P}_{MS}Kp_a}{nT} \qquad (5.42)$$

where m and n are the data segment number and coded segment number respectively. In addition, the load G is defined as the ratio between the average active UE number and the slot number. It is shown as $G = Kp_a/T$. Accordingly, the throughput TH is re-written as

$$TH = \frac{m\bar{P}_{MS}G}{n}. \qquad (5.43)$$

Based on [37], a load threshold is defined as G_{th}. If $G \leqslant G_{th}$, the collision resolution probability approaches to 1 as $T \to \infty$. Otherwise, a deep falling occurs on the collision resolution probability. The throughput of the multi-slot design is upper bounded by mG/n due to segment coding in CTSMA. For a large T, CSA is dominant to resolve the collision rather than CTSMA. Then the multi-slot design is inferior to conventional CSA due to the meaningless sacrifice on data rate. Thus this multi-slot design is suitable for short uplink contention period where the collision is mainly resolved by CTSMA rather than CSA. In addition, some IoT scenarios such as IIoT demand a short uplink contention period for the sake of latency. Thus this multi-slot design scheme has the potential to be applied in the IIoT applications.

In the following context, the collision resolution capability of the CTSMA multi-slot design is evaluated through numerical simulations. The system performance is evaluated through link level simulation on each slot of the uplink contention period. In the simulation setup, the served UE number K, the spreading factor q and the coded segment number n are fixed as $K = 64$, $q = 8$ and $n = 8$ respectively. The data packet is set to have 48 bits and the channel coding rate is 4/5 so that the channel coded bit length is $b = 60$. Moreover, AWGN channel is considered in the simulation.

At first, the BLER performance on each slot under different data segment number m in CTSMA is shown in Fig. 5.18. Besides the fixed parameters shown above, the UE active probability in this simulation is set as $p_a = 0.02$. It can be observed that error floor occurs on each of those three curves when the SNR is above 0 dB. According to the description on CTSMA in section III, the error floor is caused by colliding segments for each active UE. Since n is fixed in this simulation, the curve with smaller m has more redundancy segments so that it can resolve more colliding segments to achieve a lower error floor. Based on above, when the SNR is relatively high, the collision is dominant to impact the system performance. Meanwhile, relatively high SNR is required to achieve the reliable transmission in the mMTC system for various IIoT applications. Accordingly, collision resolution capability is the main factor to be considered.

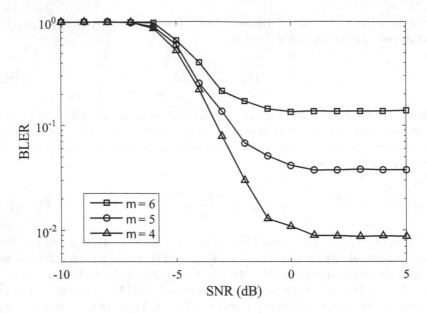

Fig. 5.18 BLER performance of CTSMA for $p_a = 0.02$

The average UE error probability on each slot with respect to the UE active probability p_a is shown in Fig. 5.19. The curves are the theoretical results derived by Eq. (5.36). The numerical results from the link level simulation for SNR = 5 dB are shown by different markers. It can be observed that the simulation results meet with the theoretical results. In addition, by manipulating the segment coding rate m/n in CTSMA, different UE error probabilities can be achieved. Therein, the transmission reliability and the sacrificing of data rate can be represented by the UE error probability and the segment coding rate respectively. For a fixed number of served UEs in a mMTC system, high reliability is achieved by CTSMA at the expense of UE data rate. Therefore, CTSMA has the flexibility to adapt different requirements of the IIoT applications.

The average UE error probability of the multi-slot design under different SIC iterations l_{max} for $p_a = 0.02$ is shown in Fig. 5.20. The lines are theoretical results derived from Eq. (5.41). The markers represent numerical simulation results. Different types of lines denote the cases of SIC iterations l_{max} and $l_{max} = 0$ indicates the case without SIC. Likewise, it can be seen that analytical derivations agree with the numerical simulations. Afterwards, for the case $m = 8$, the number of redundancy segments is 0, which indicates that segment coding in CTSMA no longer works and the threshold of the active UE number $K_d = 1$. Thus the conventional CSA scheme is represented by the curves for $m = 8$. In this figure, first it can observed that CTSMA is applied on each slot to improve the collision resolution capability of CSA due to the fact that the slot with multiple active UEs can be decoded with CTSMA. As m decreases, the slot with more active UEs can be

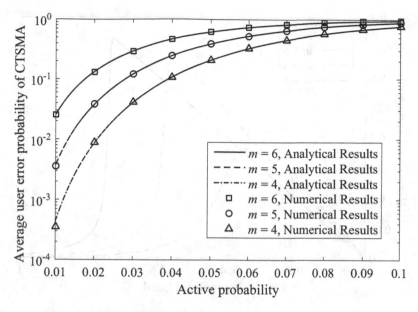

Fig. 5.19 Average UE error probability of CTSMA [32]

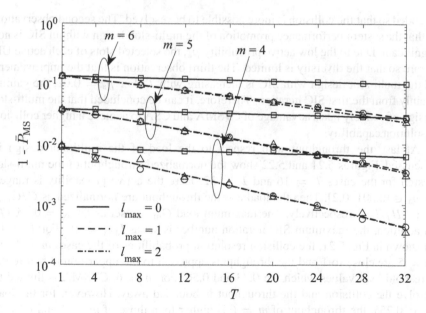

Fig. 5.20 Average UE error probability of the multi-slot design under different SIC iterations for $p_a = 0.02$

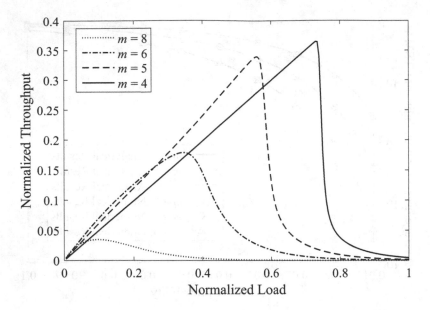

Fig. 5.21 Normalized throughput of the multi-slot design for $T = 16$

decoded so that the collision is more possible to be resolved. The second observation is that the system performance promotion of the multi-slot design without SIC is not significant. Due to the low active probability p_a, the selected slots of each active UE is rare so that the diversity is limited. The third observation is that the improvement of the multi-slot design with SIC is increasing with T. For $p_a = 0.02$, the gain is mainly from the first SIC iteration. Therefore, it can be concluded that the multi-slot design effectively takes advantage of CTSMA and CSA to achieve a higher collision resolution capability.

At last, the throughput with respect to the load of the multi-slot design is discussed. Figures 5.21 and 5.22 show the normalized throughput of the multi-slot design for the cases $T = 16$ and $T = 32$. Here the active probability is ranged as $p_a \in [0.001, 0.2]$. Then the load and the throughput are normalized as $G/G_{\max}$ and $TH/G_{\max}$ respectively. The maximum load $G_{\max}$ is set as $G_{\max} = 0.2K/T$. In addition, the maximum SIC iteration number is set as $l_{\max} = 10$. For $T = 16$ as shown in Fig. 5.21, the collision resolution probabilities of the cases $m = 4$ and $m = 5$ are close to 1 and the throughputs approach to the upper bound before the threshold load values which are 0.73 and 0.56. For $m = 6$, CTSMA is limited to resolve the collision and the throughput is bounded away. However, for the load $G < 0.255$, the throughput of $m = 6$ is higher than those of $m = 4$ and $m = 5$. It is because at low load region, CTSMA with low m is over-qualified, namely the gain from the collision resolution capability cannot compensate for the data rate sacrificing. For $T = 32$ as shown in Fig. 5.22, the throughputs of $m = 4$, $m = 5$ and $m = 6$ approach to the upper bound. The threshold values of $m = 5$ and $m = 6$

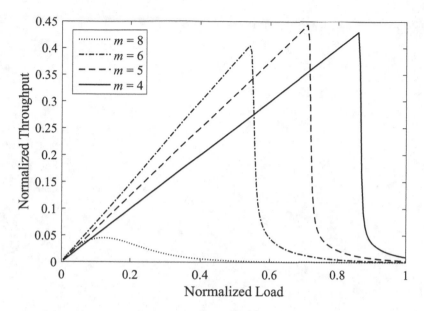

Fig. 5.22 Normalized throughput of the multi-slot design for $T = 32$

are 0.86 and 0.71. Compare to the threshold values for $T = 16$, it can be seen that the threshold value gap between different ms shrinks with the increasing T. This is due to the fact that the dominance of CTSMA to resolve the collision is decreasing with T. Based on this observation, it can be concluded that for the IIoT application which require shorter uplink contention period, the multi-slot design with small m is more suitable.

5.4 Summary

In this chapter, the novel multiple access scheme TSMA has been introduced. TSMA has three advantages. First, the number of served UEs can be much larger than the number of radio resources. Second, TSMA achieves reliable access as well as the massive connectivity, which is suitable for the future IoT applications. At last, the performance of TSMA can be controlled though manipulating corresponding parameters, which facilitates the upper layer design.

Chapter 6
Conclusion

In this book, various emerging multiple access schemes for the mMTC system have been introduced. First the characteristics and requirements of the mMTC scenario are described. For the sake of saving the enormous control signaling overheads in the small data packet transmissions, grant-free random access is regarded as a suitable random access procedure for mMTC. Nevertheless, new challenges including anonymous transmission and collision are arisen in grant-free random access. The researchers are driven to consider novel multiple access schemes. Therein, CSMUD, CSA and CTSMA are discussed. In CSMUD, sparse user activity based on sporadic transmission in mMTC is exploited to mitigate the influence of MAI. CSA enhances the collision resolution capability of ALOHA by employing SIC on multiple slots. In CTSMA, tandem spreading and segment coding are applied to confine the impact of collision to a deterministic number of segments and resolve them.

© The Author(s), under exclusive license to Springer Nature Switzerland AG 2019
F. Wang, G. Ma, *Massive Machine Type Communications*, SpringerBriefs in
Electrical and Computer Engineering, https://doi.org/10.1007/978-3-030-13574-4_6

References

1. K. Schwab, *The Fourth Industrial Revolution: What It Means, How to Respond* (World Economic Forum, Cologny-Geneva, 2016)
2. J.G. Andrews, S. Buzzi, W. Choi, S.V. Hanly, A. Lozano, A.C.K. Soong, J.C. Zhang, What will 5G be? IEEE J. Sel. Areas Commun. **32**(6), 1065–1082 (2014)
3. F. Boccardi, R.W. Heath, A. Lozano, T.L. Marzetta, P. Popovski, Five disruptive technology directions for 5G. IEEE Commun. Mag. **52**(2), 74–80 (2014)
4. P. Pirinen, A brief overview of 5G research activities, in *1st International Conference on 5G for Ubiquitous Connectivity*, Nov 2014, pp. 17–22
5. T. Wang, G. Li, J. Ding, Q. Miao, J. Li, Y. Wang, 5G spectrum: is China ready? IEEE Commun. Mag. **53**(7), 58–65 (2015)
6. L.D. Xu, W. He, S. Li, Internet of things in industries: a survey. IEEE Trans. Ind. Informat. **10**(4), 2233–2243 (2014)
7. X. Tuo, F. Wang, Y. Zhang, M. Lou, G. Ma, X. Hu, Uplink channel estimation for nonorthogonal coded access, in *2018 IEEE International Conference on Innovative Research and Development (ICIRD)*, May 2018, pp. 1–6
8. H. Shariatmadari, R. Ratasuk, S. Iraji, A. Laya, T. Taleb, R. Jantti, A. Ghosh, Machine-type communications: current status and future perspectives toward 5G systems. IEEE Commun. Mag. **53**(9), 10–17 (2015)
9. Z. Dawy, W. Saad, A. Ghosh, J.G. Andrews, E. Yaacoub, Toward massive machine type cellular communications. IEEE Wirel. Commun. **24**(1), 120–128 (2017)
10. TR 22.891, Feasibility Study on New Services and Markets Technology Enablers; Stage 1. 3GPP (2016)
11. G. Durisi, T. Koch, P. Popovski, Toward massive, ultrareliable, and low-latency wireless communication with short packets. Proc. IEEE **104**(9), 1711–1726 (2016)
12. TS 38.300, Technical Specification Group Radio Access Network; NR; NR and NG-RAN Overall Description. 3GPP (2018)
13. G. Ma, B. Ai, F. Wang, Z. Zhong, Coded tandem spreading for grant-free random access system with massive connections, in *GLOBECOM 2017 – 2017 IEEE Global Communications Conference*, Dec 2017, pp. 1–6
14. TS 36.211, Evolved Universal Terrestrial Radio Access (E-UTRA); Physical Channels and Modulation. 3GPP (2017)
15. TS 36.321, Evolved Universal Terrestrial Radio Access (E-UTRA); Medium Access Control (MAC) protocol specification. 3GPP (2013)

16. A. Azari, P. Popovski, G. Miao, C. Stefanovic, Grant-free radio access for short-packet communications over 5G networks, in *GLOBECOM 2017 – 2017 IEEE Global Communications Conference*, Dec 2017, pp. 1–7
17. N. Abramson, The aloha system: another alternative for computer communications, in *Proceedings of the November 17–19, 1970, Fall Joint Computer Conference*, ser. AFIPS'70 (Fall) (ACM, New York, 1970), pp. 281–285 [Online]. Available: https://doi.org/10.1145/1478462.1478502
18. C. Bockelmann, H.F. Schepker, A. Dekorsy, Compressive sensing based multi-user detection for machine-to-machine communication. Trans. Emerg. Telecommun. Technol. **24**(4), 389–400 (2013)
19. C. Bockelmann, N. Pratas, H. Nikopour, K. Au, T. Svensson, C. Stefanovic, P. Popovski, A. Dekorsy, Massive machine-type communications in 5G: physical and MAC-layer solutions. IEEE Commun. Mag. **54**(9), 59–65 (2016)
20. J.A. Tropp, Average-case analysis of greedy pursuit, in *Proceedings of SPIE – The International Society for Optical Engineering*, 2005, pp. 591412–591412–11
21. D. Chu, Polyphase codes with good periodic correlation properties (corresp.). IEEE Trans. Inf. Theory **18**(4), 531–532 (1972)
22. TS 36.101, User Equipment (UE) Radio Transmission and Reception. 3GPP (2007)
23. B. Ai, K. Guan, M. Rupp, T. Kurner, X. Cheng, X.F. Yin, Q. Wang, G.Y. Ma, Y. Li, L. Xiong, J.W. Ding, Future railway services-oriented mobile communications network. IEEE Commun. Mag. **53**(10), 78–85 (2015)
24. B. Ai, X. Cheng, T. Kãijrner, Z.D. Zhong, K. Guan, R.S. He, L. Xiong, D.W. Matolak, D.G. Michelson, C. Briso-Rodriguez, Challenges toward wireless communications for high-speed railway. IEEE Trans. Intell. Transp. Syst. **15**(5), 2143–2158 (2014)
25. E. Paolini, C. Stefanovic, G. Liva, P. Popovski, Coded random access: applying codes on graphs to design random access protocols. IEEE Commun. Mag. **53**(6), 144–150 (2015)
26. E. Paolini, G. Liva, M. Chiani, Coded slotted aloha: a graph-based method for uncoordinated multiple access. IEEE Trans. Inf. Theory **61**(12), 6815–6832 (2015)
27. C. Stefanovic, P. Popovski, ALOHA random access that operates as a rateless code. IEEE Trans. Commun. **61**(11), 4653–4662 (2013)
28. Y. Ji, C. Stefanovic, C. Bockelmann, A. Dekorsy, P. Popovski, Characterization of coded random access with compressive sensing based multi-user detection, in *Proceedings of the 2014 IEEE GLOBECOM*, Austin, Dec 2014, pp. 1740–1745
29. T. Richardson, R. Urbanke, *Modern Coding Theory* (Cambridge University Press, Cambridge, 2008)
30. G. Ma, B. Ai, F. Wang, Z. Zhong, Tandem spreading network-coded division multiple access. IEEE Trans. Ind. Inf. **13**(1), 390–398 (2017)
31. G. Ma, B. Ai, F. Wang, X. Chen, Z. Zhong, Z. Zhao, H. Guan, Coded tandem spreading multiple access for massive machine-type communications. IEEE Wirel. Commun. **25**(2), 75–81 (2018)
32. G. Ma, B. Ai, F. Wang, Z. Zhong, Joint design of coded tandem spreading multiple access and coded slotted aloha for massive machine-type communications. IEEE Trans. Ind. Inf. **14**(9), 4064–4071 (2018)
33. K.H. Rosen, B. Goddard, K. O'Bryant, *Elementary Number Theory and Its Applications* (Pearson/Addison Wesley, Boston, MA, 2005)
34. J.S. Plank, C. Huang, Tutorial: erasure coding for storage applications, in *Slides Presented at FAST-2013: 11th Usenix Conference on File and Storage Technologies*, San Jose, Feb 2013
35. F.J. Macwilliams, N.J.A. Sloane, *The Theory of Error-Correcting Codes. Parts I, II* (North-Holland Pub. Co, Amsterdam, Netherlands, 1977)
36. R. Singleton, Maximum distance q-ary codes. IEEE Trans. Inf. Theory **10**(2), 116–118 (1964)
37. G. Liva, Graph-based analysis and optimization of contention resolution diversity slotted aloha. IEEE Trans. Commun. **59**(2), 477–487 (2011)

Printed in the United States
By Bookmasters

Printed in the United States
By Bookmasters